ATLAS
OF
BOTANICAL
FRAGRANCE

The MIT Press
Massachusetts Institute of Technology
77 Massachusetts Avenue
Cambridge, MA 02139
mitpress.mit.edu

Originally published as *Atlas de botanique parfumée—Les complémentaires* © Flammarion, Paris, 2025

Design by Karin Doering-Froger

This book was set in Minion Variable Concept, Brandon Grotesque, and Helvetica Neue by the MIT Press. Printed and bound in Spain.

Library of Congress Cataloging-in-Publication Data is available.

ISBN: 978-0-262-05176-7

10   9   8   7   6   5   4   3   2   1

EU Authorised Representative: Easy Access System Europe, Mustamäe tee 50, 10621 Tallinn, Estonia | Email: gpsr.requests@easproject.com

Jean-Claude ELLENA

illustrated by Karin DOERING-FROGER
translated by Erik BUTLER

ATLAS
OF
BOTANICAL
FRAGRANCE

The MIT Press
Cambridge, Massachusetts
London, England

Our taste for spiritual matters is accompanied by passionate curiosity about the circumstances that have given rise to them.
Paul Valéry

Each perfumer has their own way of classifying scents. Usually, they are grouped by olfactory similarities based on years of experience. This book takes an approach that is historical, poetic, geographical, and botanical. It describes fragrances in terms of the emotional bond I hold with each one and, since it is an atlas, according to their origin; to complete the picture, discussion includes the part of the plant from which scents are obtained. Among other things, we will learn that star anise, long the most expensive spice in the world, replaced absinthe, the "green fairy," to bring forth pastis, Marseille's most popular drink. Tar from white birch, which is the source of leather odors, was used by Neanderthals as a glue for affixing arrowheads. Now that plants have been recognized as possessing a form of intelligence, we know that the leaves of the acacia tree represent the result of co-evolution: the plant developed clever mechanisms to avoid being eaten by giraffes, which the animal in turn circumvented.

CONTENTS

It is not against nature to be curious. It is in our nature to be so.
Carlo Rovelli, *Seven Brief Lessons on Physics*

INTRODUCTION

A maker of perfume enamored of fragrances and the stories behind them, I wrote the first *Atlas of Perfumed Botany* in 2020. Few publications or studies in the field of botany treat scents. Scientists are not unaware of the smells that plants emit, but even the simplest flower expresses itself in hundreds of fragrant molecules. It takes an empirical approach to pick out, from this tangle of smells, those particles that would enable one to classify them in full. Accordingly, research on plants has privileged sight, an imperious faculty that blinds us to the other four senses.

The olfactory sense, which is the first one to develop *in utero*—vision is the last— did not really receive scientific attention until the 1970s. This is when the perfume industry developed techniques for capturing odors much in the same way that sound is recorded, but with different equipment, called *Head Space.*[1] Perfumers joined forces with botanists and chemists to scour the planet and collect impalpable traces, first on the ground and then on the forest canopy, without removing plants from their surroundings. This approach, which was economical and sparing, sought to show that the perfumer's craft followed nature and, at the same time, deserved the status of a science. Because captured and reconstituted scents don't always match those of plants in the wild closely enough, the new millennium has witnessed the emergence of new professions, for example that of *sourcer* —a person who finds raw material not for eating but for smelling (and at times both). The job entails traveling the globe and tracking down new fragrances to be cultivated in a sustainable, ethical way that respects the people and places encountered. Today, perfumers have never had so many natural extracts at their disposal. These substances, which enable them to innovate and tell new stories, are the stuff of dreams.

---

1. This technology is primarily used to capture the odors emitted by plants (flowers, fruit, etc.) *in situ*. Substances caught by an absorbent filter are analyzed and identified using chromatography and mass spectrometry in the laboratory, then recreated.

A hard, compact, solid substance that forms the roots, stems, and branches of trees and shrubs.
Littré Dictionary

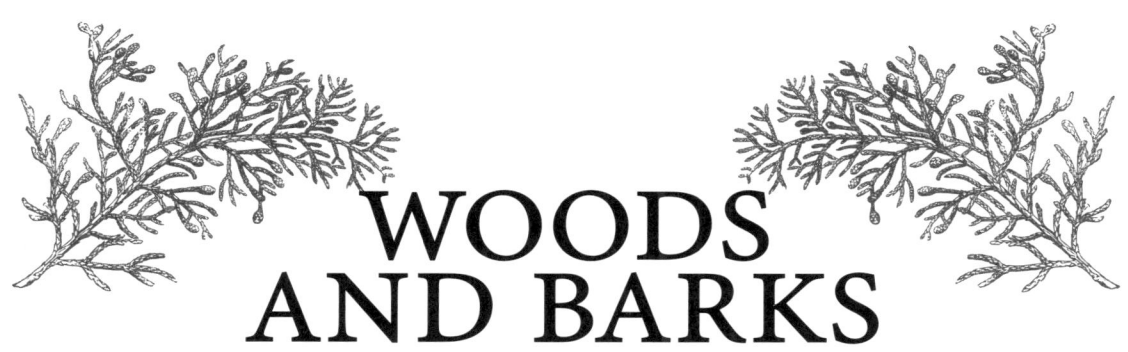

WOODS
AND BARKS

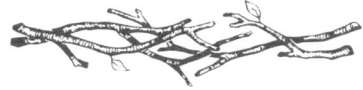

The dictionary definition of wood tells us little about what it's made of and nothing about our relationship to it. For most people, wood is an everyday object that has lost all its luster. We no longer know—or have forgotten—that the chairs we inherited from our parents or grandparents are made of oak, the table and cupboard of walnut, or the bookcase of cherry. We pay no mind that the garden chairs are teak, and the handles on our garden tools ash; even the crate we used to bring fruit and vegetables home from the grocer this morning is made of poplar or pine. There's no denying that companies selling assembled furniture and do-it-yourself kits have not gone easy on this resource; even when lumber is "certified European," people don't know what it's called or where it's from, especially when material from different places is thrown together.

Wood regains its dignity and pedigree when the framework of Notre-Dame Cathedral is rebuilt—or, more simply, when a craftsperson restores a piece of furniture, a cooper repairs barrels, a luthier makes an instrument, an architect designs a house, or a perfumer works with essential oils, the very soul of wood.

ROSEWOOD

Aniba rosaeodora

What is rare sometimes lies far, far away. The journey to the land of rosewood is unexciting as far as Lima. Air France offers daily service from Paris to the Peruvian capital. The flight covers 12,000 kilometers and takes twelve and a half hours to complete; there's a six-hour time difference. Once you've arrived, it's a two-hour flight by prop plane to Iquitos, a town in the north built partly on the river and inaccessible by road. Iquitos is well worth the trip for its picturesqueness. Then you board a *piragua* and embark on two days of travel up the Amazon. Sights along the way include sloths and monkeys lying in the trees or hanging from branches, the legendary scarlet macaw in its regal plumage, and a few deceptively sleepy alligators on the riverbank. At long last there's the rosewood forest, the final destination.

Then you board a piragua and embark on two days of travel up the Amazon.

Our story does not date back far. It starts with resistance mounted by a handful of landowners in the area. When the government planned to clear part of the Amazon forest in order to set up industry in the town of Tamshiacu, old-growth rosewood was harvested illegally and savagely, and it almost disappeared. The overexploitation led to the species receiving protection in appendix 2 of CITES (*Convention on International Trade in Endangered Species of Wild Fauna and Flora*), which permits only regulated commerce. The local population decided to grow rosewood in accordance with the new rules; in this way, economic support was provided for the 6,000 inhabitants of Tamshiacu. The extraordinary tale calls to mind Jean Giono's widely translated story, "The Man Who Planted Trees"—except that it was not just one person planting individual trees but a hundred or so planting a whole forest. The first difficulty was that parcels of land were unregistered. The families involved had long known each other and respected property limits, but as far as the government was concerned, the terrain had to be surveyed and zoned for use. The requisite steps were taken. During the long bureaucratic interlude, nurseries were established, cuttings were made, and eventually 60,000 young trees sprang forth. Using GPS technology, the growers recorded the date each tree was planted. To promote general harmony in the forest, other kinds of trees were also introduced. It's the same with trees as it is with people: too much of the same stifles growth. After seven or eight years, the trees were ready to be harvested—on the condition that as many as were felled would be replanted. The wood was then cut up and chipped with an axe or machete for distillation.

Native to South America, rosewood grows in Brazil, Peru, Amazonia, and French Guiana. An evergreen with small flowers clustered in axillary panicles, it reaches an average height of twelve meters. Rosewood started to be prized in the eighteenth century under the reign of Louis XV, when it was used for the marquetry and veneer of chests, tables, and cupboards, in keeping with the demands of aristocratic clients. The appeal of this hard, compact wood is its pink color, which can vary between lighter and darker shades.

Two centuries later, a 90 percent essential oil was made by distilling shavings that don't smell like roses so much as linalool, its main component. The fresh, flowery, and slightly peppery scent resembles that of the essential oil of coriander seeds. The main component is identical. In decreasing order of importance, it also occurs in extracts of bergamot, lavender, lavandin, orange blossom, jasmine, and rose.

Rosewood has been grown for perfumery since 1997, and its essence is a regulated commodity. Today, most linalool is synthetic and not extracted from a natural source, which protects the environment. Nevertheless, Europe classifies this aromatic compound as one of the twenty-six allergens requiring oversight. Present in the most common natural essences to various decrees, it has come to be used in any number of fragrances on the market. Rosewood essence is one of the components of Chanel's Bois des Îles (1926), which I love unconditionally, Dior's Miss Dior (2019), which is very different from the 1947 version, and, most recently, Hermès's H24 (2021), an eau de toilette for men in which it features prominently.

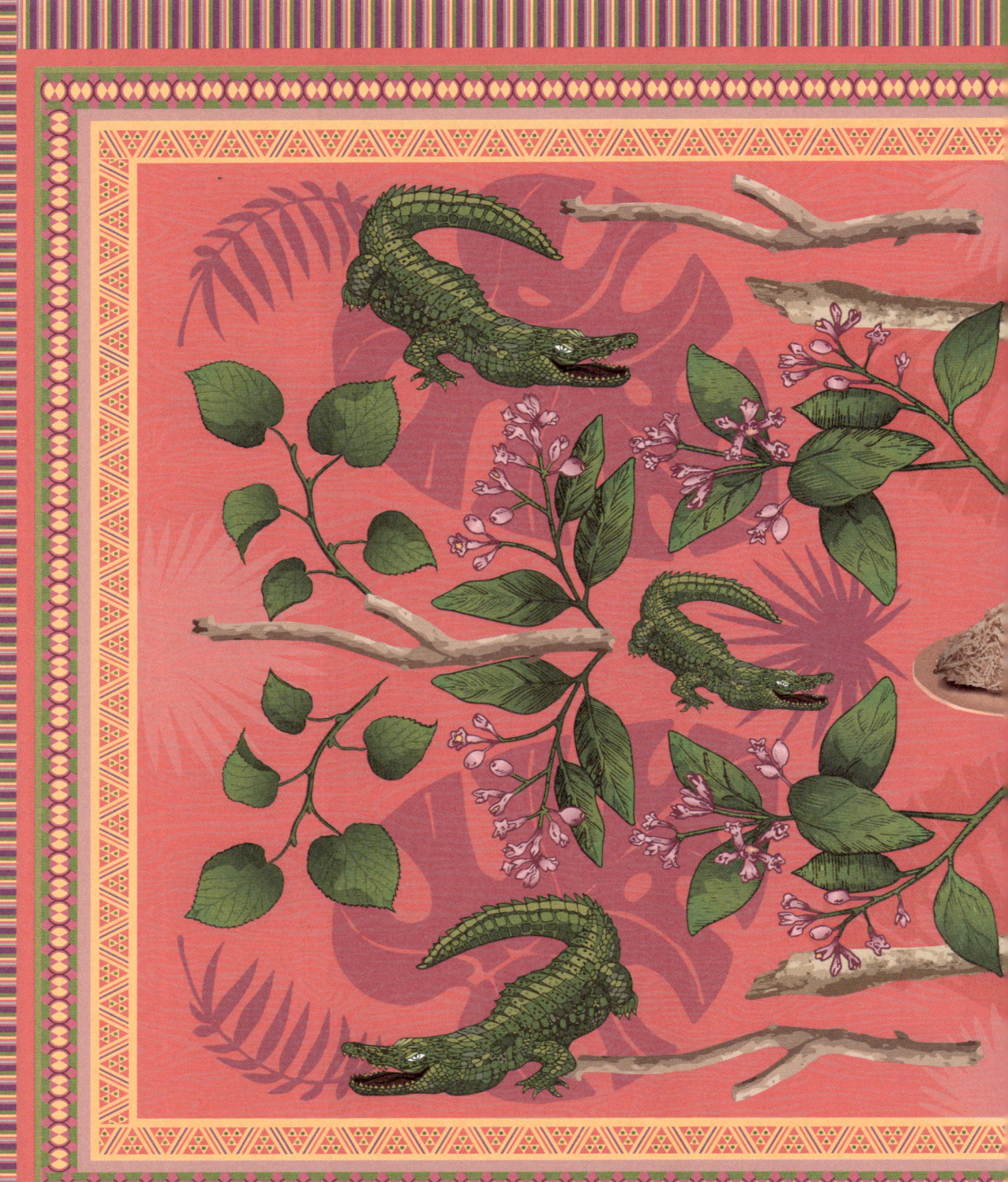

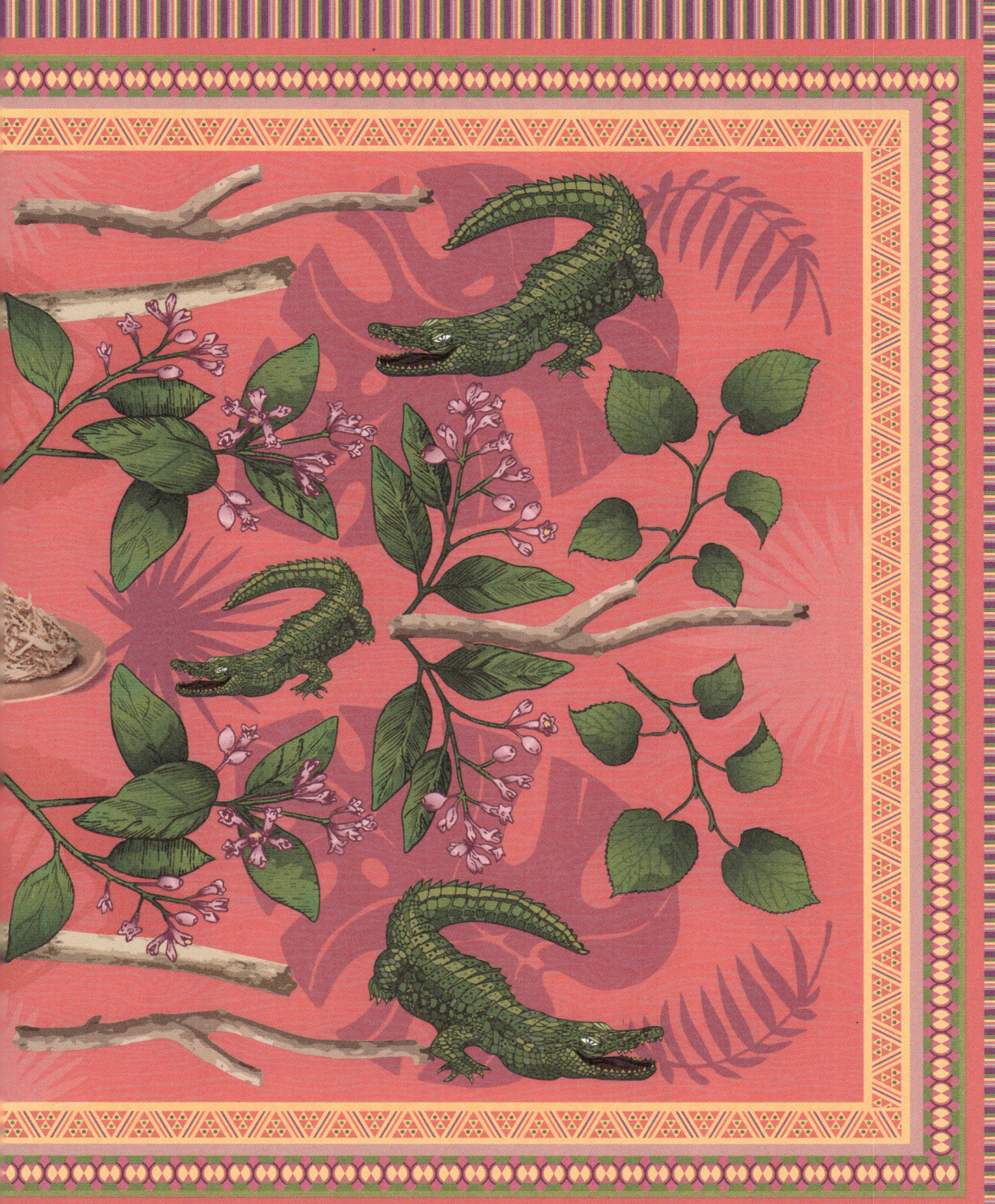

WHITE BIRCH

Betula papyrifera

A hundred years ago in 1924, on the night of June 20, violent storms inundated the Paris region, particularly Arcueil. Winter floods were still fresh in residents' minds, including those of 1910. In spite of the awful weather, an optimistic mood prevailed at the Théâtre des Champs-Élysées. Here, the Ballets Russes were presenting *Le Train Bleu*, an *opérette dansée* by the celebrated Bronislava Nijinska. Darius Milhaud composed the music, Jean Cocteau wrote the libretto, Coco Chanel designed the costumes, and Pablo Picasso made the stage curtain. The ballet took its name from a luxury express line between Calais and the Mediterranean that was rechristened in 1922; starting in the late nineteenth century, it had connected England and the sea to the south, dancing along the steep cliffs and bright gulfs. Even though the press did not take a favorable view of the production, it caused quite a stir. Soon, Chanel's Cuir de Russie came out. It was easy to see a connection between the name of the perfume and Mademoiselle Chanel's bold patronage of the arts. But the name could have had another origin, too: her conquests included Grand Duke Dmitri Pavlovich of Russia, cousin to Tsar Nicholas II.

The connecting thread in this story is a tree, the white birch, which is the symbol of Russia; its tar is used for tanning and its oil for perfumes. Cuir de Russie by Chanel would attain renown for providing the archetypal leather note in perfumery. The same year—although the date of creation is controversial—another perfumer released a scent with a similar note, Knize Ten. Knize was founded in Vienna in 1858 by couturier John Knize, the Austrian royal family's official supplier in Paris. The man behind the fragrance, Vincent Roubert, became head perfumer for François Coty at the same time; he had already been working for him under the table at the Grasse workshops of Antoine Chiris, sole supplier to the firm. Ernest Beau, the creator of Cuir de Russie, worked exclusively for Chanel. The two houses stood in competition; each spied on the other and weighed the merits of its frenemy's creations. History doesn't say which perfume came first and inspired the other; all belongs to the realm of conjecture.

For an artist, norms, fashions, and conventions are there to be subverted. Accordingly, in 1925, perfumer Jacques Guerlain used tar oil to provide an animal note in Shalimar; its proximity to the scent of castoreum is plain. In 1998, it was used for its smoky, woody note in Déclaration by Cartier. Tar oil's peaty qualities were also featured in Épices marine

(2013).[2] Scents are like colors: We make them say what we want them to say.

In prehistoric times, Neanderthals discovered that heating birch bark under hermetically sealed conditions produces tar and ash (this is known as dry distillation). They used the tar as an adhesive; ancient arrowheads bear traces of where it was used for fastening tips to wooden shafts. They also valued it as a disinfectant; because of its phenol-rich composition, it penetrates and sanitizes the skin by cauterization. (In turn, much later in history, ashes would play a major role in the production of printing ink.) From the eighteenth to the early twentieth century, the West provided a major market for Russian leather, which was made into fine bookbindings, coach interiors, trunks, and boots for coachmen and soldiers. The secret was a substance that nourished the hides after tanning, a mixture of birch tar oil and other fats such as castor oil and fish oil. This compound strengthened the grain, hardened the surface, and rendered the leather highly water-resistant. Dephenolizing tar oil by washing it in an alkaline medium produced the essential oil of birch so important for perfumes.

A large tree, the white birch can reach twenty-five meters in height. Native to North America and Russia, it reproduces by seed and by propagation, when its roots bring forth shoots of new plants. The serrated leaves are simply shaped. The tree makes fruit in the form of samaras clustered in catkins that delight a wide range of birds including partridges, nuthatches, and titmice.

White birch is also known as canoe birch. Native Americans used the bark to cover their boats, as well as for their homes. The wood is soft, firm, and knot-free, making it suitable for all kinds of small objects pertaining to food and eating: chopsticks, sticks for lollipops and ice cream, toothpicks, and wooden spoons. Finally, its sap is used to make sugar and a refreshing wine that has been appreciated for centuries in Russia and Northern Europe.

---

2. In the Hermessence collection.

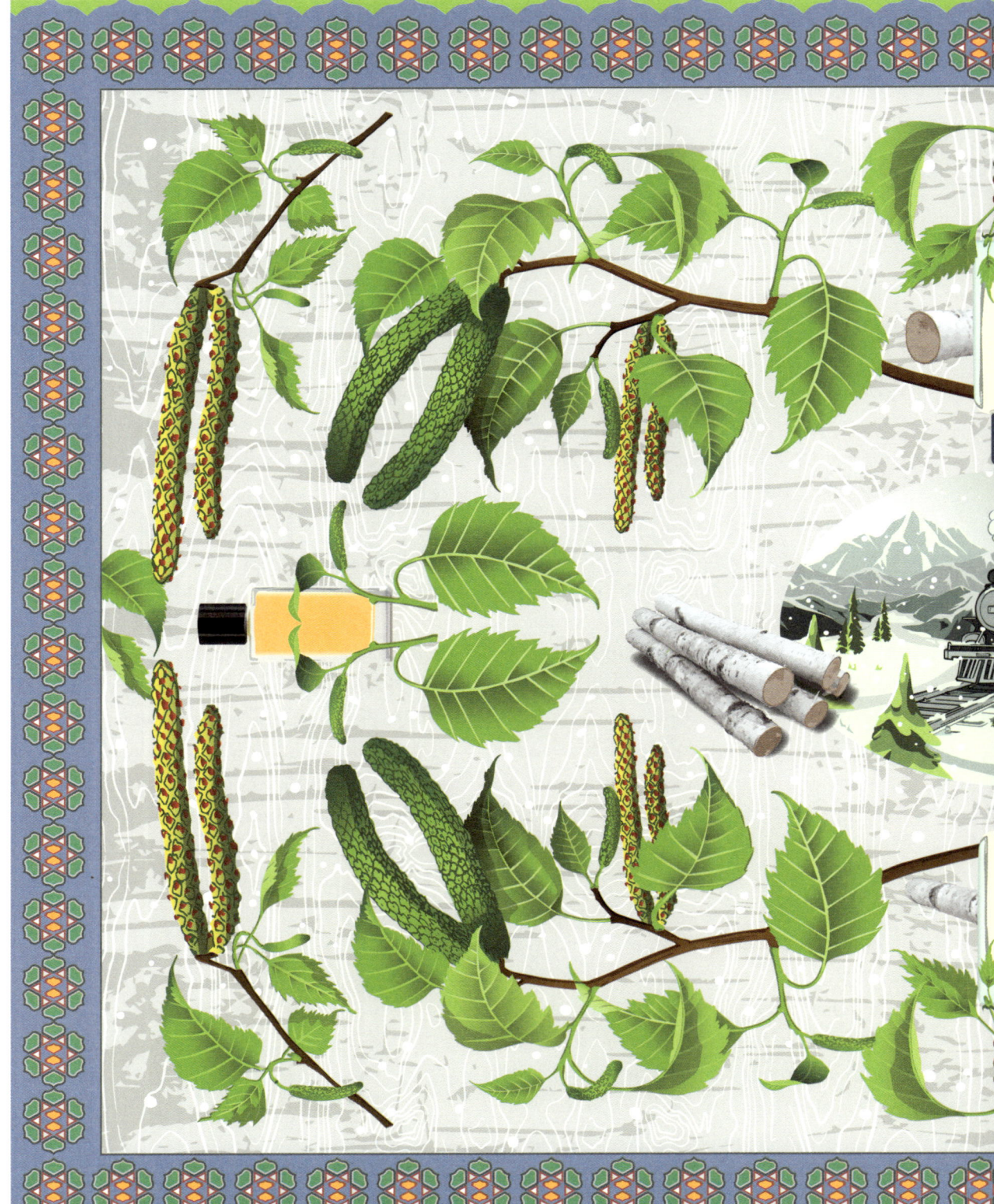

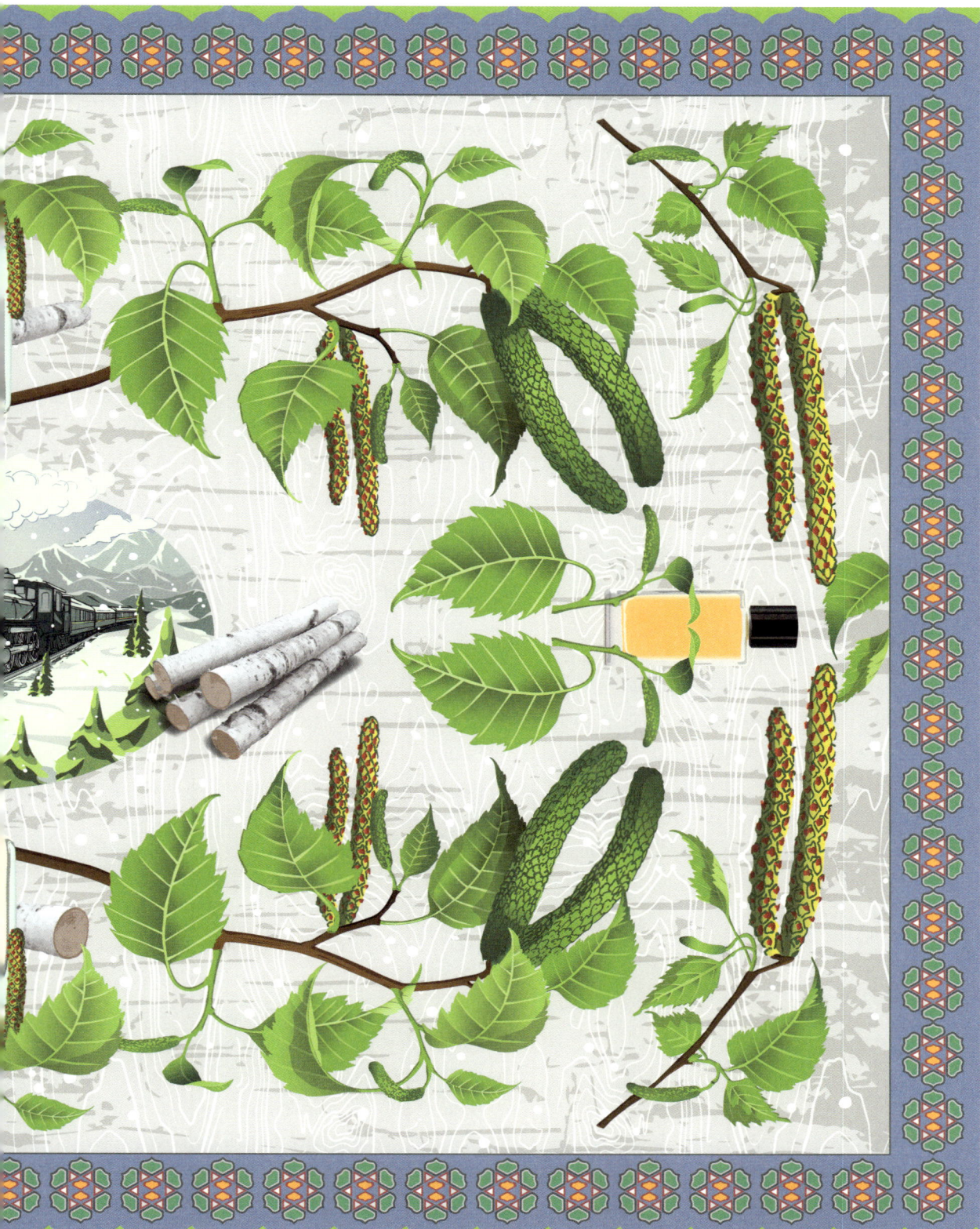

CYPRESS

Cupressus sempervirens

As a child, I accompanied my grandmother when she went to pick "the flower," her name for jasmine. The harvest began around July 15—today, it starts a month earlier—and lasted for three months, sometimes four. To reach the property, we took a narrow path, a ribbon of dry, white earth worn by the footsteps of people out for a walk and laborers on the way to work. She led the way, and I followed. For a hundred meters or so, we skirted a rampart of cypress trees that stood shoulder to shoulder and smelled of carrot root and turpentine. This arboreal screen hid what I thought was an immense château. My curiosity was whetted, but all I could see was a low gray wall.

The Mediterranean or Italian cypress comes from Central Asia, and not from the Mediterranean—as you might think if you go by what you see in Spain, France, Italy,

Greece, and elsewhere, too. It is our good fortune that the plant immigrated when it did and shapes the landscape today. Slender in form, it can reach twenty to twenty-five meters in height. Its evergreen foliage lends it rare elegance and, as an old belief would have it, betokens proximity to the gods themselves. Along with olive trees, cypresses shape the gardens of Villa Adriana in Tivoli, and with palm trees they structure the gardens of the Alhambra in Granada. Cypresses punctuate the Tuscan landscape and define the dramatic perspectives of the long roads leading to villas. Just recall the last scene in Ridley Scott's *Gladiator*: the dying hero opens a large wooden door and walks up a magnificent cypress-lined avenue to paradise, where he will be reunited with his loved ones. In the less populous parts of Provence, it is a custom, a tradition, and a courtesy to plant three cypress trees at one's house; just one would be a matter of happy accident. That way, the voyager, after a long day's journey, can spot the place where he'll be well received and find food and drink.

The tree's life expectancy of 500 years symbolizes immortality. For this reason, it is frequently planted at the entrance to cemeteries on the southern side—and not in the north, as it hates the cold. The cypress should

> For a hundred meters or so, we skirted a rampart of cypress trees that stood shoulder to shoulder and smelled of carrot root and turpentine.

not be confused with the yew, which is also a conifer and highly prized in parks and at cemeteries beyond the Loire. Cypresses were first planted in Provence to protect crops from the mistral, the cold, powerful wind that originates in the Alps, winds down the Rhône valley, and picks up speed as it rushes to meet its mistress, the Mediterranean. Just

Its evergreen foliage lends it rare elegance and betokens proximity to the gods themselves

get on the A6 freeway or the N7 after Valence, put on a song by Charles Trenet, and you'll see them alongside the poplars, marking out the fields like soldiers standing at attention. And if you can't tell these trees apart: The poplars are old and have put on weight; the cypresses are young, tall, and wiry.

Cypress distillation is a recent development for perfumery. The first trials were carried out in Alpes-de-Haute-Provence in 1904 with the bushy twigs obtained from annual pruning. The essential oil smells of juniper berries, resin, and pine, with which

it is sometimes treated. After two days of evaporation, the essential oil leaves an ambergris-like odor on the blotting paper strip. Chemists have tried to isolate this precious molecule, but to no avail. Most scents are composites, and do not depend on a single element, as one might hope. Cypress is used in the composition of Calèche by Hermès (1961). In this perfume, Guy Robert transformed Madame Rochas—which he created the previous year for the couturier Marcel Rochas—by adding vetiver and cypress; these sleek and slightly austere scents bring out the masculine quality of saddlery in a fragrance that is otherwise wholly feminine. Hermès again picked up cypress in 1998 for Rocabar, a manly eau de toilette, and in 2021 the perfumer Armani released, from his private collection, Cyprès Pantelleria, an ode to the Mediterranean.

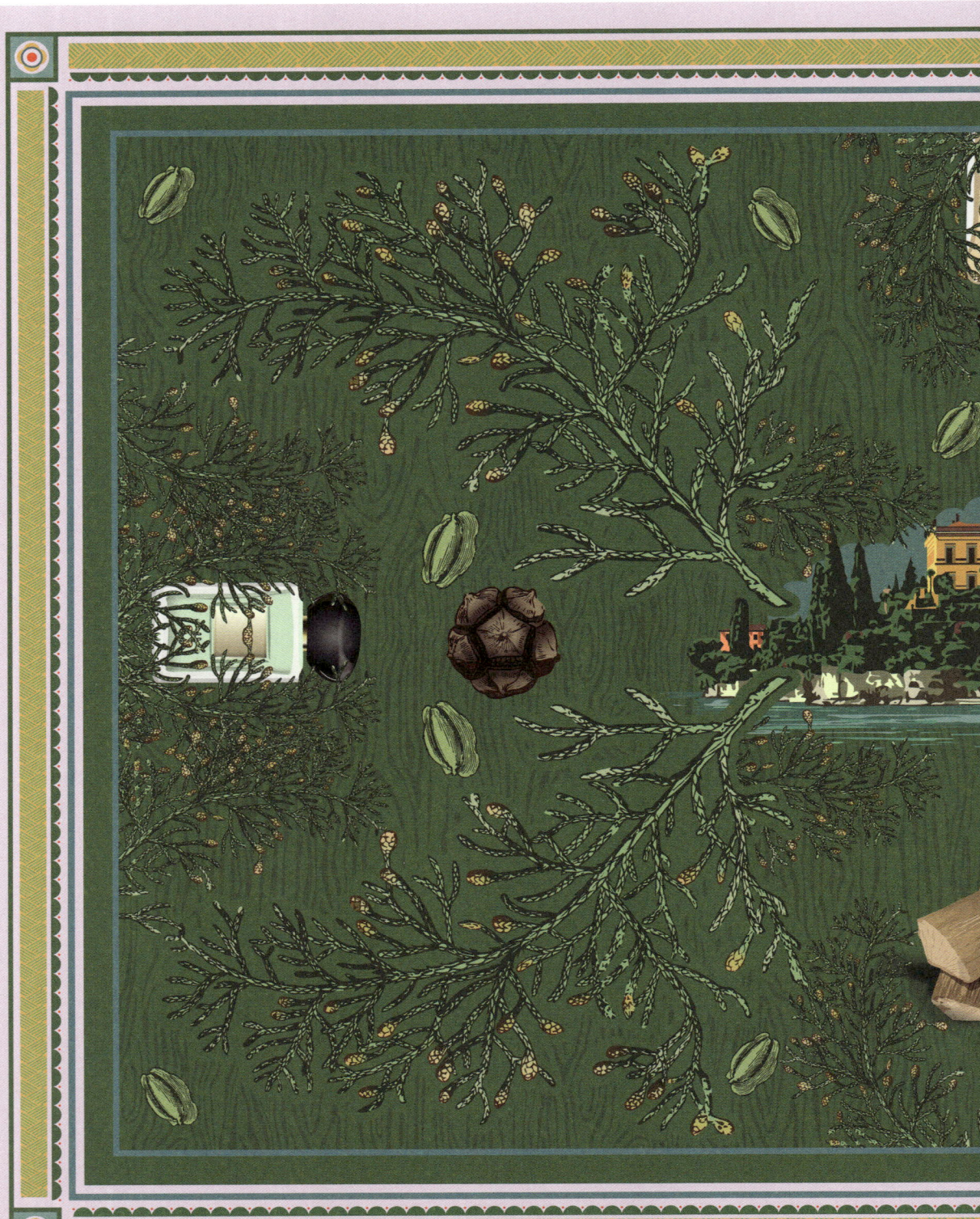

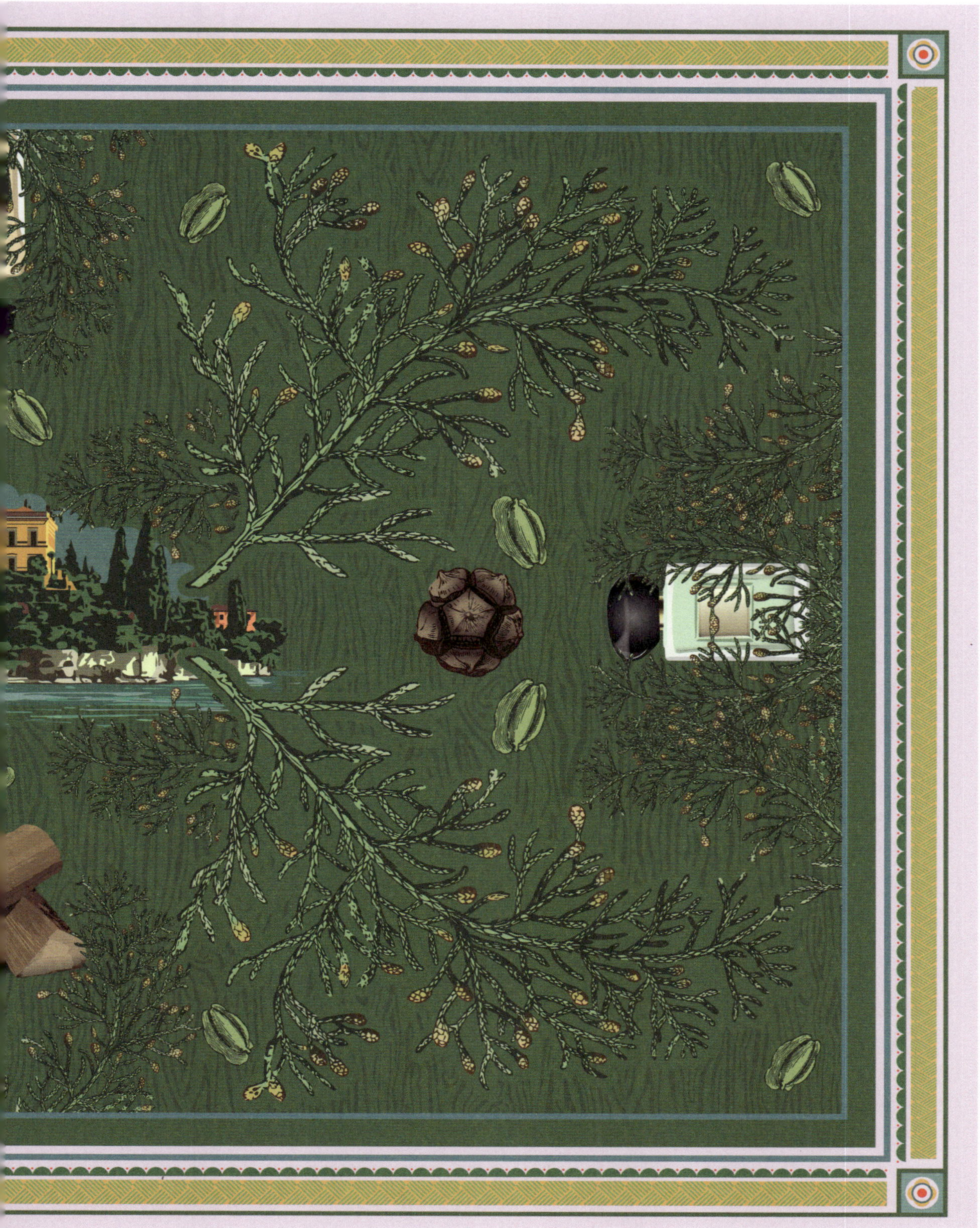

BALSAM FIR

—•—

Abies balsamea

From the 1970s on, the boom in tourist travel opened the world to new flavors and scents. For French perfumers, the winds blew westward. In America, they discovered techniques and methods, based on the study of consumer needs and market structures, for expanding perfume sales; this is what would come to be known as "marketing." They also discovered a number of new scents, including balsam fir absolute. It debuted in 1965 in Aramis by Aramis, a highly concentrated men's fragrance that met with great success in the United States. Fir balsam is likewise found in Paco Rabanne's famous Paco Rabanne Pour Homme (1973) and Guy Laroche's Drakkar Noir (1982). In 1992, Pasha by Cartier gave this fragrance a particularly luxurious image.

The balsam fir is native to North America. It looks like the Christmas tree of children's dreams. Taller than other firs and spruces, it can reach up to twenty meters in height, and its conical shape lends it fitting elegance. Tougher than its relatives, it sheds its needles late in the season; the latter are particularly flat. To thrive in the forest, it prefers to keep to itself and enjoys sunny spaces. The tree, which likes cold climates and moist soils, grows at altitudes of up to 1,700 meters, and it can live to be 150 years old. The bark of young balsam firs has vesicles or glandular reservoirs containing a resin from which a turpentine called Canada balsam is extracted; it serves to glue optical lenses and repair cracks in glass such as windshields. Native American tribes were the first to use this resource. Medicinal applications included antiseptic ointments, poultices, and chewing gum to ward off colds. Fir needles were also made into herbal teas and used to fill pillows; the scent was thought to prevent illness.

Legends surrounding this tree with a hundred uses and a thousand virtues are legion. It is said that when Jacques Cartier landed in the Gulf of St. Lawrence in 1534, a scurvy epidemic was decimating his crew. The unlucky captain had the good fortune to cross paths with the son of a Native American chief who knew the virtues of *annedda*, the name for balsam fir, and some of his men were saved. In thanks, Cartier named the territory Canada, a name that likely derives from the Huron and Iroquois word *kanata*, or "village." He also brought seeds of what he had christened the "tree of life" back to Francis I. Balsam fir enjoyed extraordinary popularity in its day; over time, alas, it and its wondrous qualities have faded into oblivion.

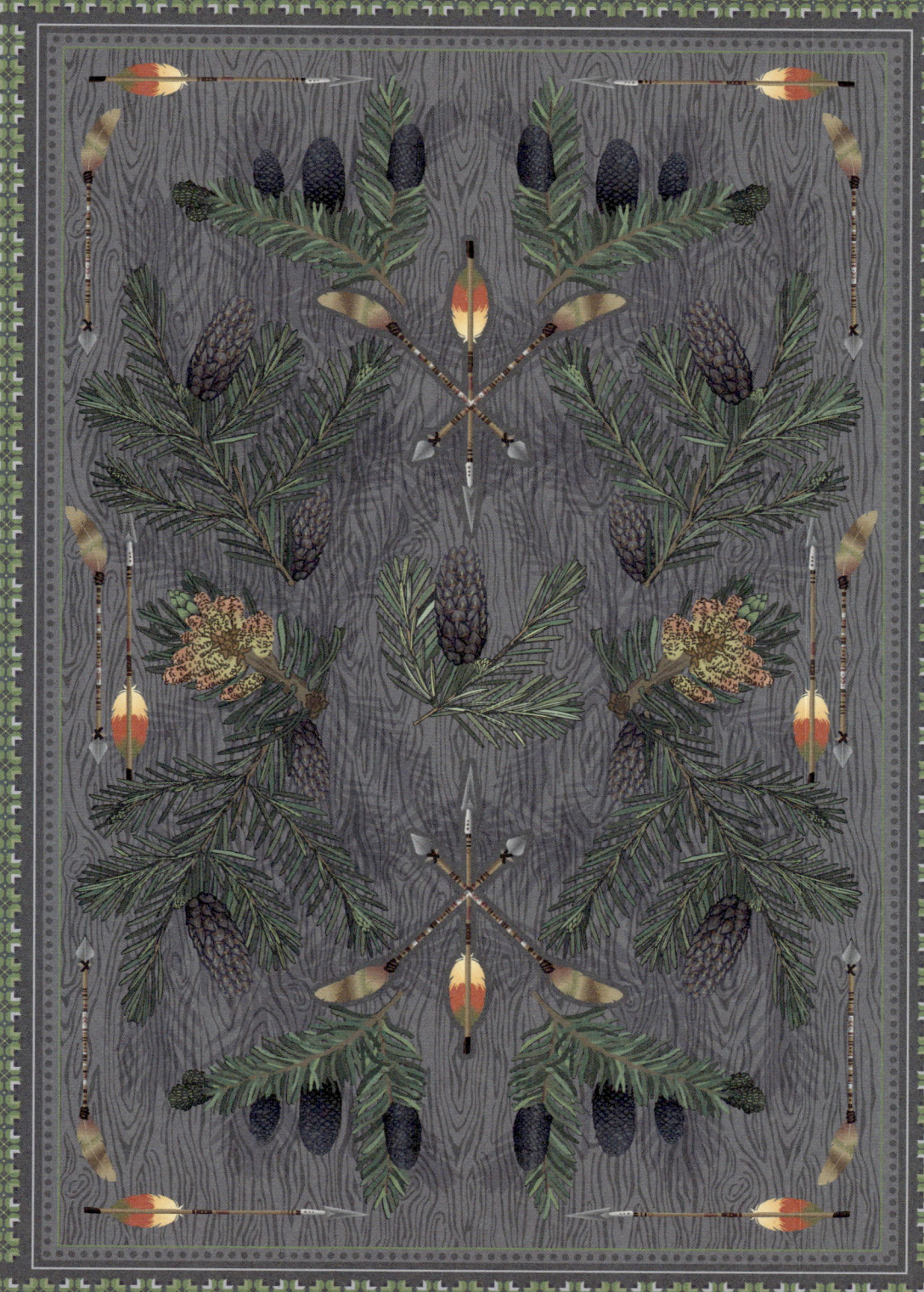

GAIAC

—•—

Guaiacum officinale

"Strike!" she yelled, leaping up to express her joy. In a single stroke, she had knocked down all ten pins in lane seven of the bowling alley. Did she know that the ball she had rented was made of guaiac wood, the densest and hardest in the world, and that this same ball was heavier than water?

Gaiac, known informally as ironwood, is used to manufacture turbine bearings, propellers, and pulleys for ships. In the days of pirates and privateers, it was made into wooden legs when limbs were lost in whale hunts—a practice that came to an end when the fairy known as Electricity rendered whale-oil lamps obsolete. Modernization has its advantages. The most famous leg belonged to the cunning and brutal Long John Silver in Robert Louis Stevenson's *Treasure Island*. With his parrot Flint on his shoulder, he long epitomized the pirate in the world's collective imagination. Now, Johnny Depp's Jack Sparrow, from *Pirates of the Caribbean*, has usurped his place for younger generations. To each their own.

This tree is small, not reaching more than ten meters in height. It grows slowly, and its trunk and twisted branches are evergreen.

Gaiac is distinguished by greenish-gray bark that, when it peels off, yields a pattern calling to mind the spots of a leopard— like a plane tree. The foliage is dense and lends the tree a compact quality. Grouped in bunches and blooming profusely, the flowers have a mauve-blue color that withers and whitens over time. Native to the Caribbean, gaiac is found in the Windward and Leeward Islands of the West Indies, as well as in Colombia, Venezuela, Guyana, and Paraguay. Its gray-green essential oil has a thick consistency and a rose-like scent with a smoky undertone reminiscent of dried prunes. Fragrances in which it is featured include Van Cleef & Arpels Pour Homme by Van Cleef & Arpels (1978), One Man Show by Jacques Bogart (1980)—to excess— and Paris by Yves Saint-Laurent (1983); in the latter case, the American perfumer Sophia Grojsman used it to supply rose notes. Another example is Rose Poivrée by Different Company (2000). Most recently, it has appeared in Oud Laqué by Bains Guerbois (2022). Difficulties in sourcing its essential oil have made it a rare and expensive commodity.

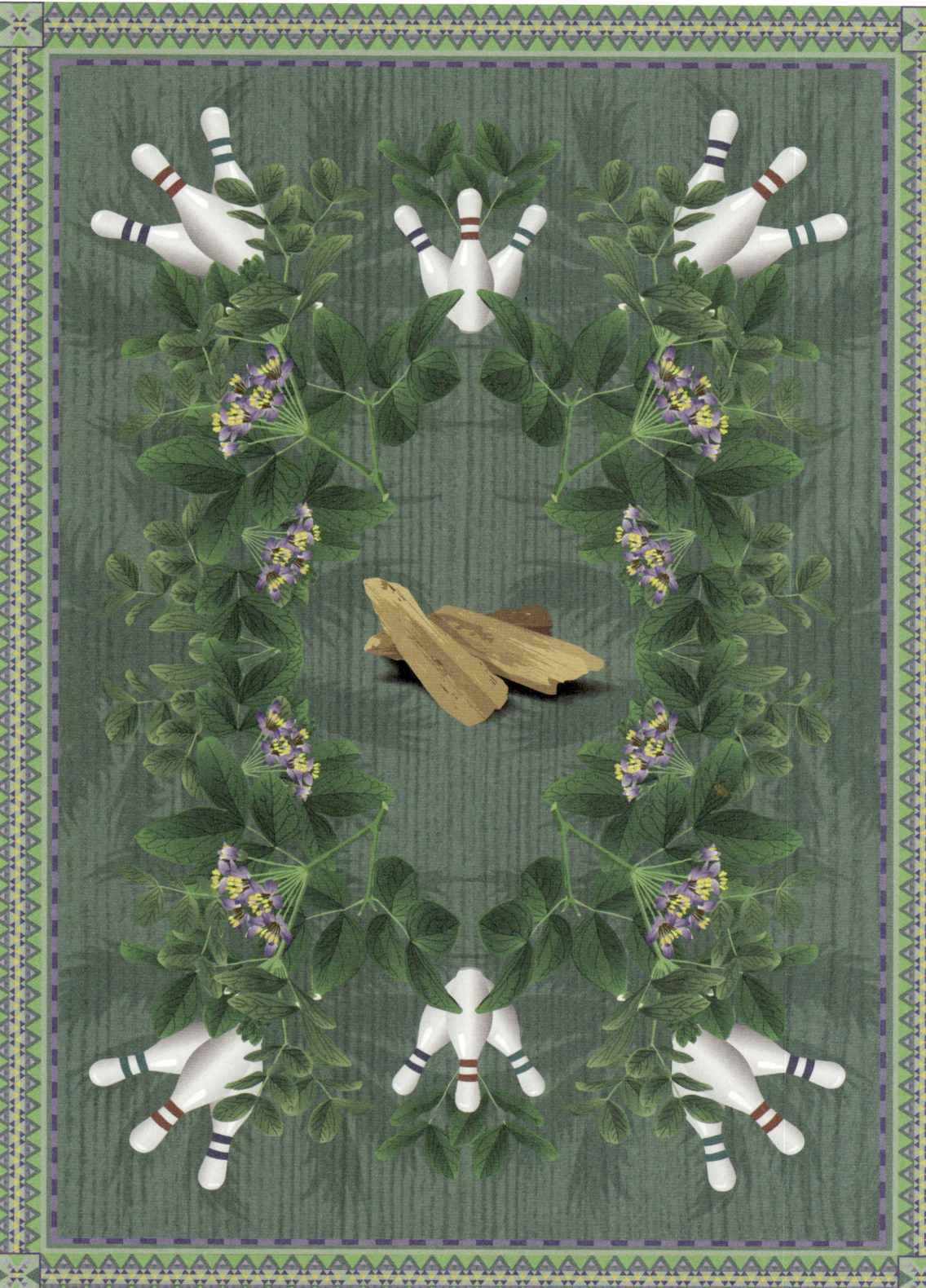

OUD

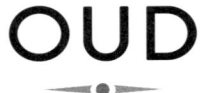

Aquilaria crassna

I've heard it enough times to make my head spin. "Try some oud! The Middle East and America are crazy about it. It's all the rage!" This isn't the lute that is widely played in the Arab world, whose slightly mournful sound I rather like. We're talking about an extract from a wood: *Aquilaria crassna*.

There isn't a luxury line—Dior, Cartier, Tom Ford, Armani, Vuitton, Hermès—that doesn't have an "oud" on the market. If I disapprove, it's not by my nature, but because all fashions inevitably go out of fashion. Plus, I don't like doing what everybody else is doing. It's already been done—why repeat things? People tell me that oud "moves." Sales

> Buying a perfume should be a matter of desire, of the love one is ready to give in tribute.

aren't the point. Being wanted is. That's the definition of luxury for me. Selling something takes pitches, lines, spiels, and song and dance. Buying a perfume should be a matter of desire, of the love one is ready to give in tribute.

Oud, agarwood, aloeswood, and gharuwood are found in the tropical forests of Southeast Asia. There are at least twenty species that produce a fragrant natural resin called calambac in response to injury suffered by the tree (wounds, fire) or because of biological attack (insects, bacteria, and fungi). The name of the wood is used interchangeably with the resin it produces. *Aquilaria crassna* provides the most famous calambac or oud.

These woods have been esteemed in China, India, and the Middle East since time immemorial—further back than I've been able to determine. In the sixth century before the Common Era, the previously nomadic Nabataeans are said to have seized Petra (Jordan) and adopted settled ways of life. In turn, they made the city an important center for trade in frankincense and myrrh from southern Arabia (which represented their primary source of wealth), as well as sandalwood and oud from India. The latter was and still is credited with medicinal virtues. It counts as a kind of panacea, a universal remedy.

Demand is so high and overexploitation so great that, since 1970, various calambac-producing species have gone extinct in India and South-East Asia. Worried nations have asked for the trees to be listed in appendix 2 of CITES, which allows trade only under regulation. To keep up with the ever-expanding market, and to avoid illegal or uncontrolled sales (which can also drive prices down), essential oil companies are now setting up partnerships with Laos, Thailand, Cambodia, and Indonesia. Here, plantations inoculate the trees with disease. The fast-growing tree is cut in its seventh or eighth year, then sorted into brown or black wood and white wood. Calambac from brown or black wood (which is more infected) is distilled to produce oud or agarwood oil. Calambac from the less contaminated white woods yields boya oil. Although quality and cost vary in keeping with provenance, the price per kilogram can reach 25,000 euros for oud, the most expensive essential oil in the perfume industry, and 14,000 euros for boya.

I sometimes marvel that off-putting scents are often the most expensive. This is true of ambergris, animal musk, civet, and castoreum. Oud smells not like rotting wood but like sweat: the adrenaline of wood under attack. A major component is caprylic acid, which smells like goat's milk (from which it gets its name). As a perfumer, I like animal scents because they are also human

and forge a bond. In the absence of animal notes, I have used costus (*Saussurea costus*), a root with the odor of greasy hair and sweat that is used in Indian cuisine; now, however, this extract has been banned in perfumery because of its allergenic properties.

Oud failed to move me, so I wondered about its appeal. Was it the price, the rarity, the supposed medicinal properties? Over time, I came to believe that the fascination it exercises comes from its smell, after all. And since I believe that scents need a story, I pictured myself in a Bedouin tent in one of the Arabian deserts, seated on a large carpet and living on what my few sheep, goats, and camels could provide. I tried to rediscover the scents of a time when the point of life was not to hurry. Among other things, perfume means finding one's place.

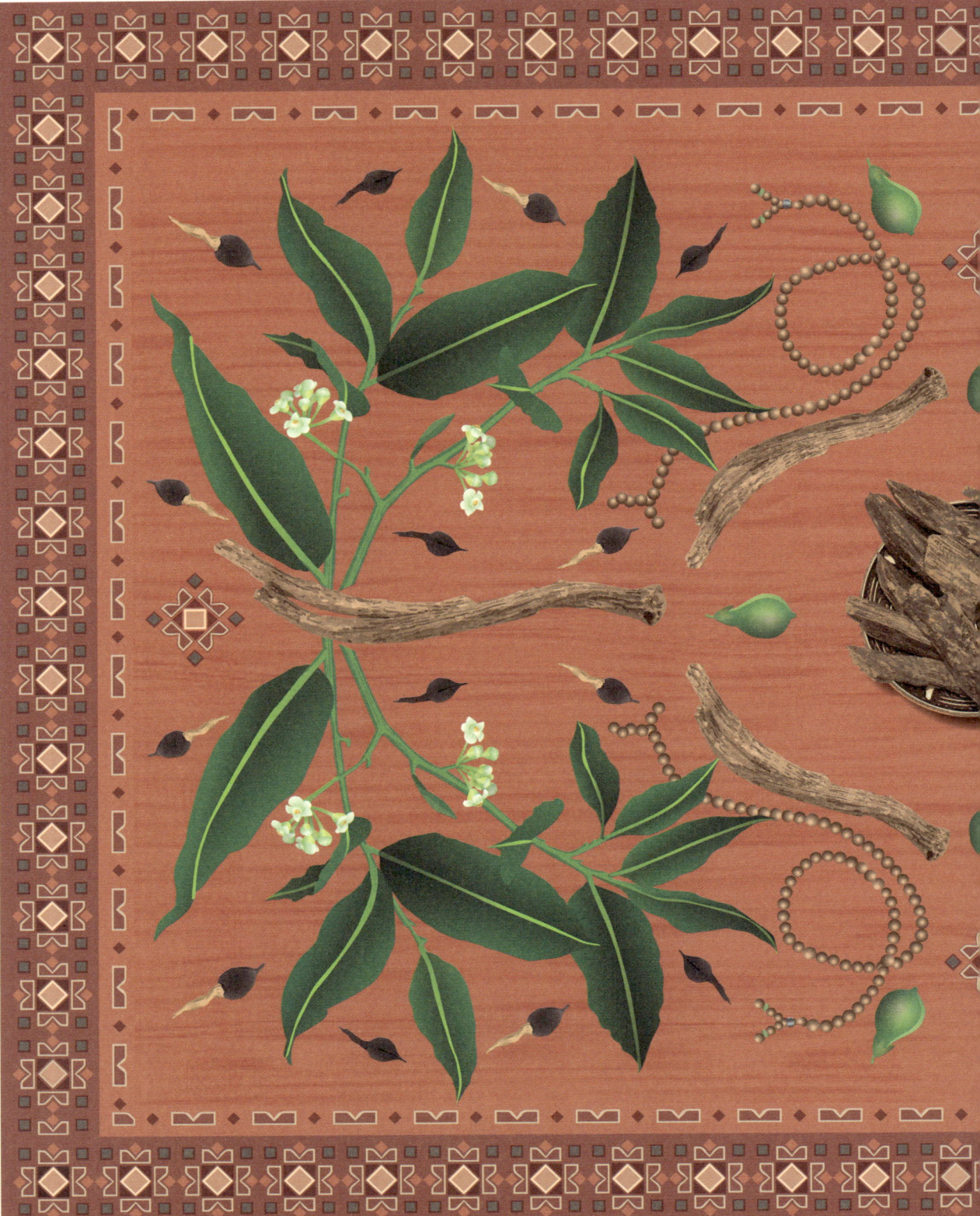

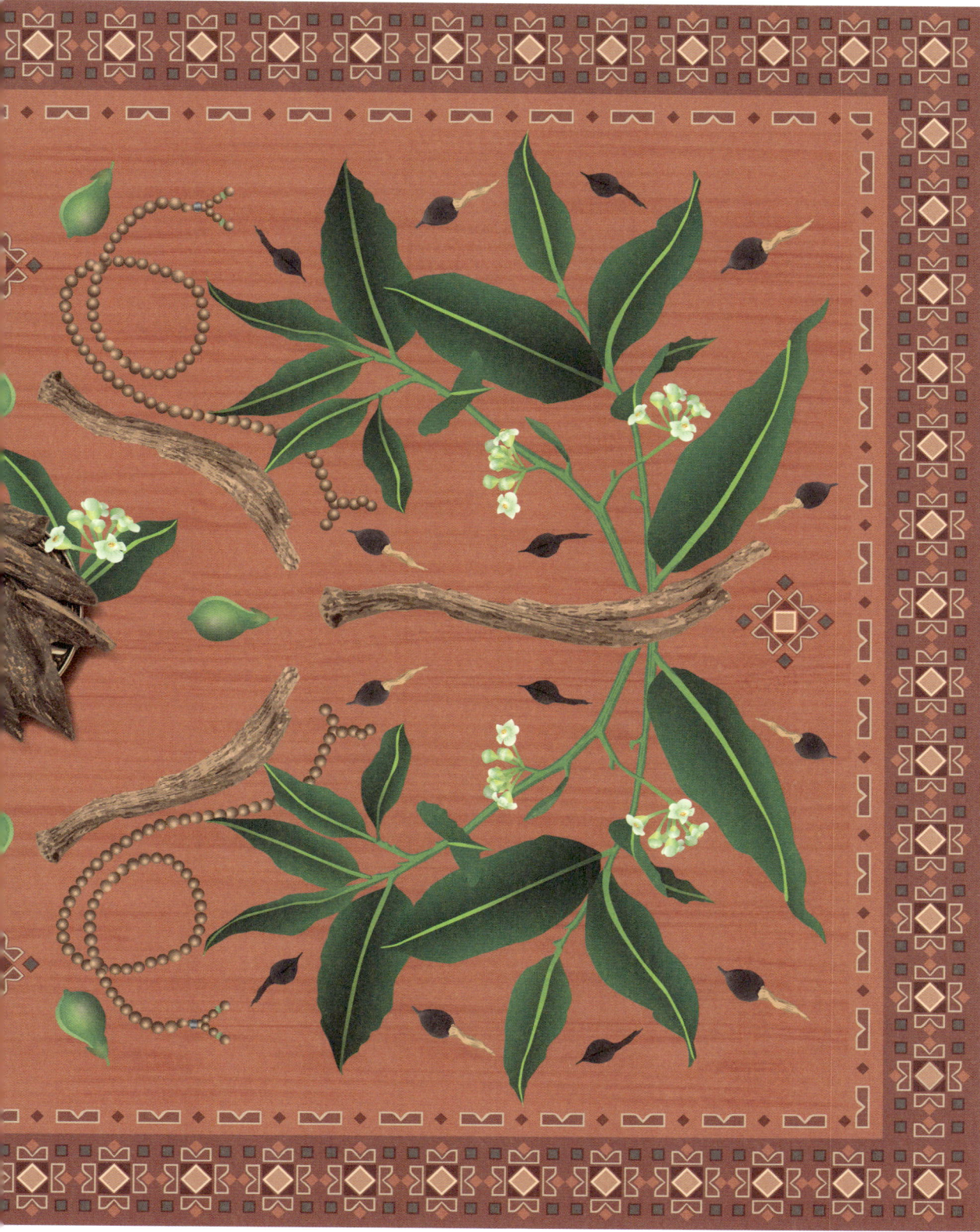

Thin, flat, usually green part of the plant, sprouting
from stems and branches.
Littré Dictionary

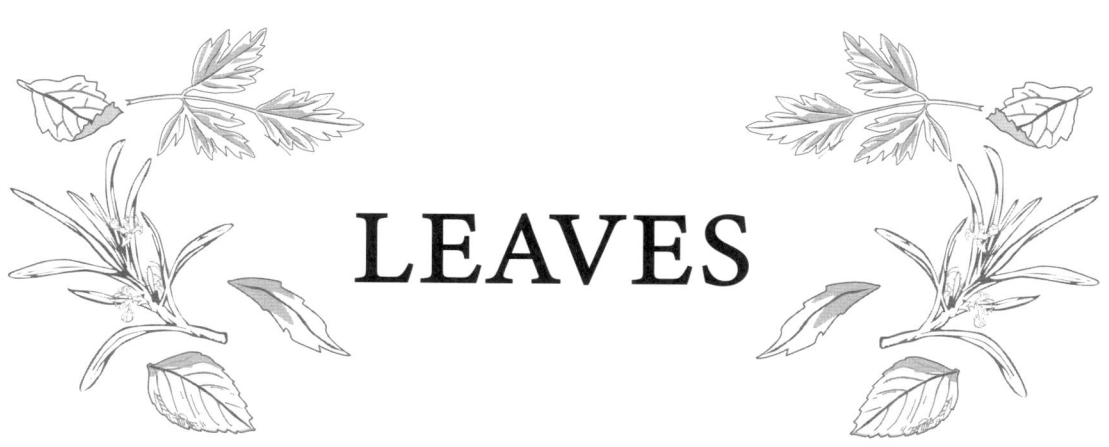

LEAVES

There's an exquisite milieu, the world of cooking, where leaves and herbs are particularly sought after. Wild or cultivated, they enable us to tell stories, rediscover childhood memories, and offer up nature itself on a plate. I still remember visiting the garden of Michel Bras, a Michelin-starred chef in Aubrac. It was early morning, and the dew was making pearls on the grass. My host put on green boots and invited me to join him. I could hardly keep up; he picked herbs as fast as he walked. He plucked and passed me a leaf of *Cedronella canariensis*, which smelled like cooked onions to him. I replied by mentioning Indian *Asa foetida*, a gum-resin with a similar scent, with which he was familiar. Then he gave me a leaf of *Mertensia maritima*, which is crunchy and slightly bluish and has an oyster-like taste; I was bewildered to find the sea in my mouth. He handed me a *shiso* leaf. I loved the scent. Words failed me until I struck upon the cross between the smell of an inner tube and that of mint. My guide found the description bizarre. We passed an assortment of mints and basil, and he recited the name of each one, pointing out costmary, or mint geranium, and lovage, which tastes strongly of bouillon cube. Then we went into one of the greenhouses, where I tasted some leaves of Vietnamese *rau-ram*, which reminded me of the smell of tobacco juice; he didn't like them, either. I was delighted, and even happier, to find that our approaches, our passion for taste, smells, herbs, and leaves, were similar. Jean Sulpice, another Michelin-starred chef, has confirmed the connections we made. He writes: "Mint brings freshness, tarragon aniseed flavors, and chives sweetness; parsley completes the snail's return to its world of herbs."[3] Perfumers, who use mint, tarragon, and parsley but not chives (yet), could say the same.

3. *L'Assiette sauvage, 50 recettes aux herbes et aux fleurs* (Le Cherche-Midi, 2015).

DAVANA

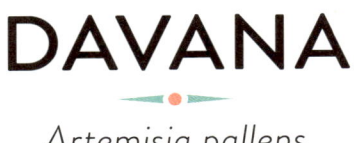

Artemisia pallens

With Hermès, a firm named after the god of communication and voyages, I traveled the world. My trip to India made a lasting impression. To speak of India is to speak of color, sound, touch, and taste—and, for a perfumer, smells. Everything is more forceful, more fascinating, heavier. I cannot think of another country where scents are so assertive and overwhelming. They attend life day and night from birth to death, and even beyond. They include davana, a fragile artemisia of the same genus as mugwort, tarragon, and wormwood. This little-known plant, which belongs to the *Asteraceae* family, grows in China, India, Pakistan, Afghanistan, the United States, and even France, although it is not cultivated there. In India, it plays an important role in offerings to the gods and is highly prized. An annual native to China, it thrives on the temperate slopes of the Himalayas and in the Kashmir valley. India is the sole country where it is grown commercially. Harvest usually begins in late February or the first week of March. The plants are harvested when half of them have reached the flowering stage; they are then cut at the base, shade-dried for forty-eight hours, and steam-distilled to obtain the maximum amount of essential oil.

There are two grades of davana essential oil. One comes from Karnataka, Maharashtra, and Kerala, that is, the southwestern region along the Malabar coast; it is my favorite. The other comes from Tamil Nadu at the southeastern tip of the subcontinent, in other words, the side on the Bay of Bengal. Annual production yields eight tons of essential oil. Davana is used mainly for flavoring pastries, beverages, and tobacco. Its syrupy scent calls to mind aged spirits such as cognac and armagnac, as well as dried fruits such as apricots and prunes. Perfumers particularly appreciate the fragrance and use it in trace amounts. Its main component is davanone, a lactone, followed by nerol and geraniol; the latter are found in the scent of roses, with which it is often combined. Perfumes featuring davana essential oil include Coriandre by Jean Couturier (1973), Élixir des Merveilles by Hermès (2006), Davana & Vanille Bourbon by 100 Bon (2017), and Honeysuckle & Davana by Jo Malone (2018).

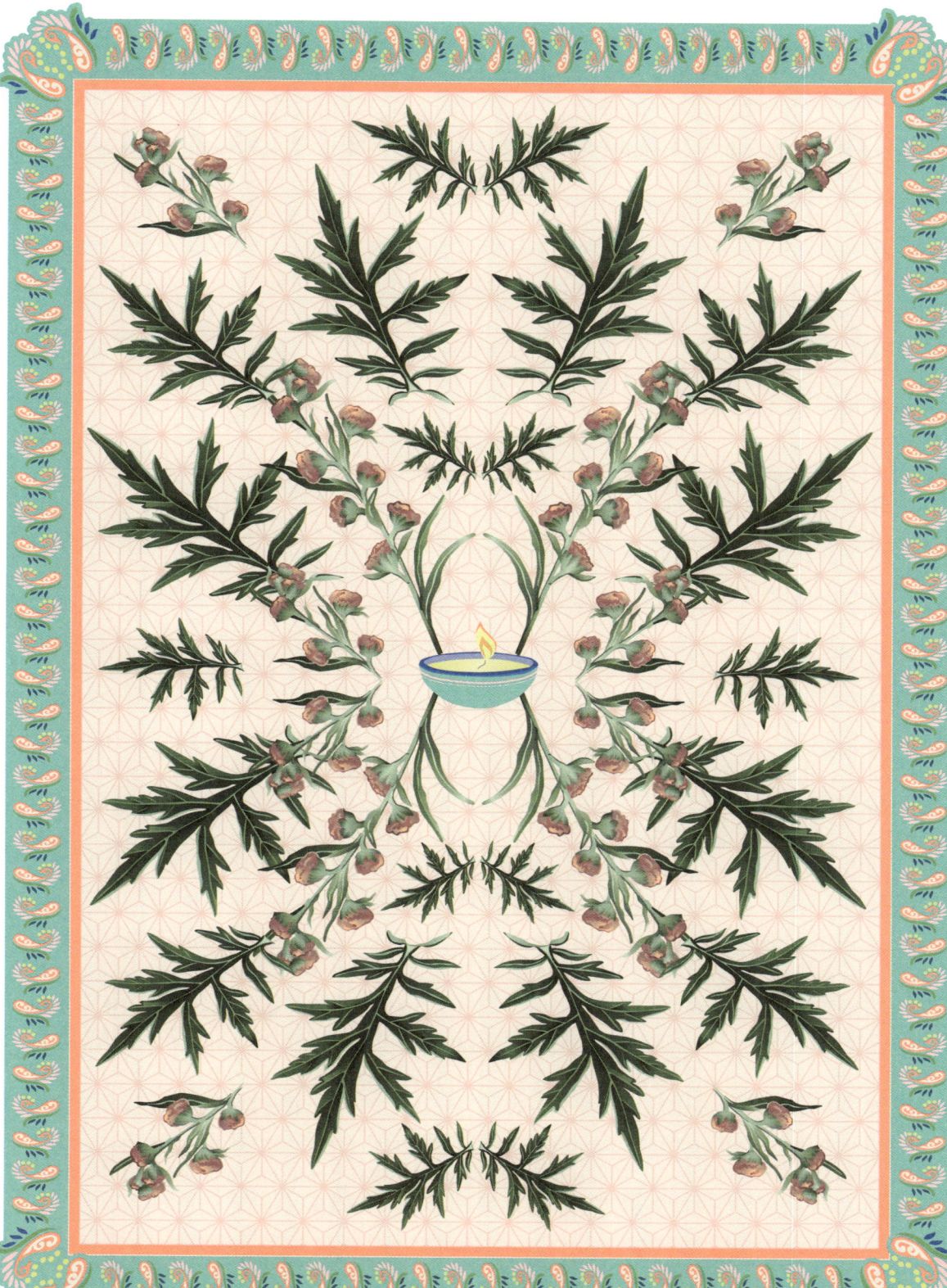

MASTIC OR LENTISQUE

Pistacia lentiscus

For a long time, galbanum provided the green note in perfumery. It asserted itself to excess in Vent Vert, a fragrance crafted by Germaine Cellier for Pierre Balmain. This perfume debuted at the Cannes Film Festival in autumn 1947, when the new palace hall (which is not the one we know today) first opened.

Since the 1970s, perfumers have turned to a new green note: mastic or *Pistacia lentiscus*, with its scent of crumpled leaves. Unlike galbanum, which is an umbelliferous plant, mastic is an evergreen shrub with beautiful green leaves that are shiny on top and dull on the underside; in wintertime, they take on an attractive crimson hue. In the past, the leaves were used in tanning and gave color to leather from the Grasse region in France. The plant, which generally reaches one to three meters in height, grows abundantly along the Mediterranean coast and is not to be confused with the deciduous terebinth. A marvel in the garden, it requires little care, and its foliage smells heavenly. The earliest mention of the shrub occurs in the works of the philosopher and botanist Theophrastus. Roman writer and naturalist Pliny the Elder

detailed its medicinal uses. Greek physician and pharmacologist Dioscorides did the same and attributed numerous virtues to its leaves, fruit, wood, and resin. Among other things, he recommended the sap for styling women's eyelashes and as a chewing gum to freshen breath—especially the *chia* variety from the Aegean island of Chios. For 2,000 years, Chios has been producing the finest gum; it has boasted the AOC label since 1997, hence its nickname of "Mastic Island." Greece, Turkey, Syria, Iraq, and Iran consume mastic in abundance, using it in pastries, candies (Turkish delight), liqueurs, cosmetics, incense, scented pastilles, and to manufacture varnishes for the fine arts.

It was not until the twentieth century that perfumers began experimenting with distillation and extraction with volatile solvents to obtain the lentisque absolute used in perfumes; they also used an essence from Morocco. Mastic is found in Paco Rabanne Pour Homme (1973), Eau de Campagne by Sisley (1976), Un Jardin en Méditerranée by Hermès (2004), and, more recently, (Untitled) by Martin Margiela (2010).

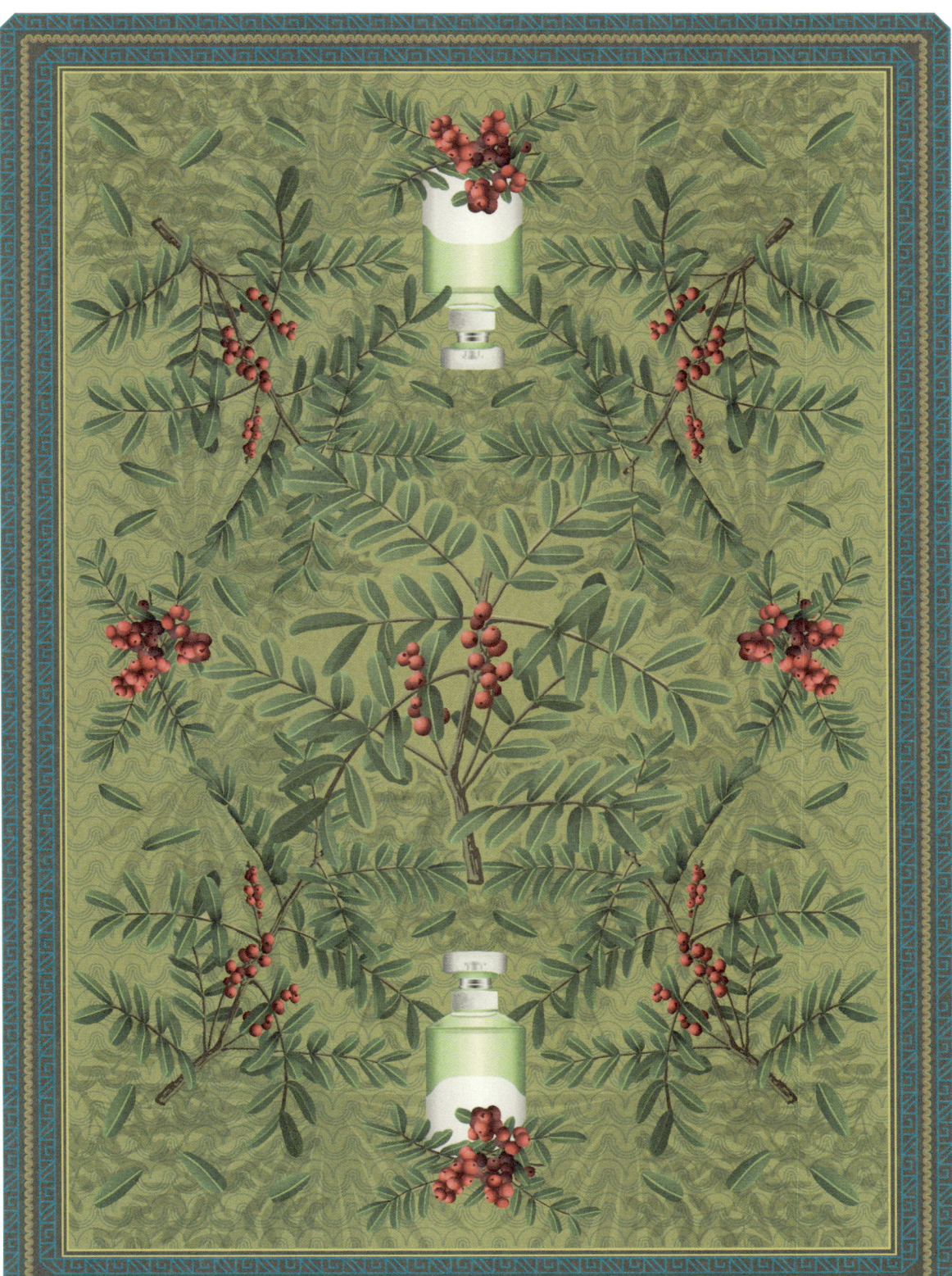

LOVAGE

Levisticum officinale

Groix, a small and charming island, lies thirty-nine kilometers by land from Belle-Île-en-Mer, where the painter Claude Monet famously went to paint. At the end of the nineteenth century, Brittany was fashionable—as it is now becoming again—and for Monet, Belle-Île-en-Mer epitomized its charms. Groix, which is less well known, is nicknamed the "Island of Garnets" for the color of the sand on its beaches. Lovage leaves are grown, dried, ground, and sold here as a local specialty. In his book *L'Assiette sauvage* (*The Wild Plate*), the Michelin-starred chef of Val-Thorens, Jean Sulpice, recommends combining lovage, or mountain parsley as it is also known, with the iodized flavor of cockles. For the perfumer, lovage essence, whether distilled from roots or leaves, provides the saltiest scent available. The reader might be surprised that salt has a smell. Salt is first and foremost a taste; that's its defining trait. But it's no coincidence that celery salt is made from lovage seeds. To my knowledge, no essence has ever been extracted from them.

A perennial member of the *Apiaceae* family native to Iran, lovage is the ancestor of all forms of celery and turnip. The plant grows at altitudes below 1,800 meters in the Pyrenees, the Alps, Central Europe, and the Caucasus. It is especially prevalent in Hungary, Bulgaria, Romania, and Moldavia; here, under the name *leustean*, it is used to enhance the taste of *ciorbà*, a popular soup.

Although it has disappeared from our kitchen gardens for the time being, lovage deserves a place alongside parsley, sage, tarragon, and basil. Ten times more powerful in taste and smell than celery, its roots and leaves call to mind the scent of bouillon cubes. The signature component is sotolon or 3-hydroxy-4,5-dimethylfuran-one. This is the molecule that gives yellow Jura wine its distinctive character and contributes to the taste of old rums and sake, as well as certain whiskies. Few molecules make such a lasting olfactory impression on their own.

Perfumery is an art of illusion, let us recall. Magic happens when lovage essence is used to accentuate vetiver, as in the very beautiful Eau de Vétyver by Givenchy (1959)—and, even more, Sel de Vétiver by Different Company (2006) and Vetivera by Couvent Maison de Parfums (2022).

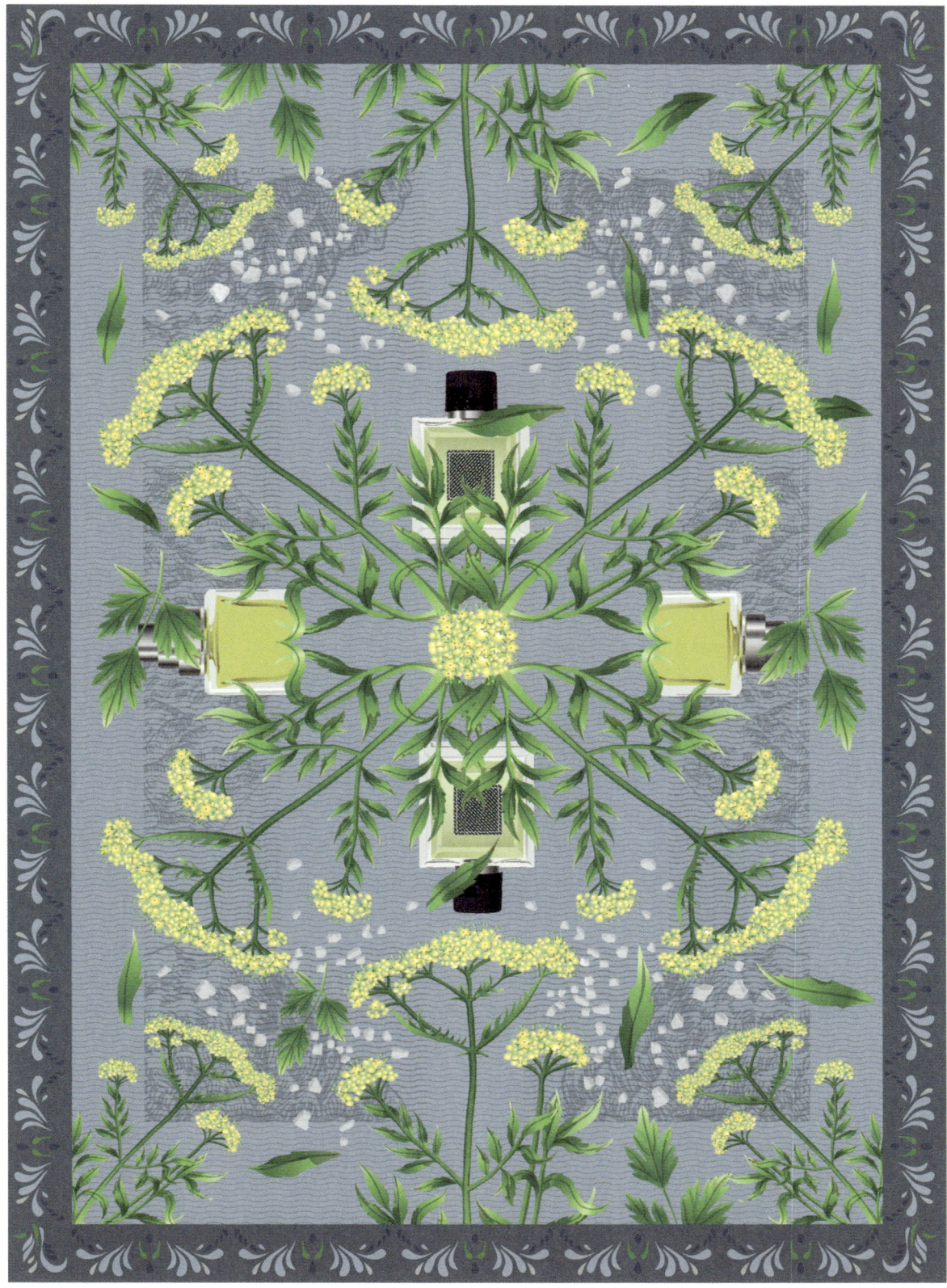

MINT

— • —

Levisticum officinale

These herbaceous plants are grown all over the world; their taste and fragrance have played a role in daily life since the dawn of time. Egyptians, Greeks, and Romans knew several species of mint, but since ancient accounts concern only how they were used, and not what they were, our picture of them is imperfect. Of the many species that exist, only four are important for perfumery.

Although I'm no botanist, I am curious. I find that the various kinds of mint look very similar; it's not always easy to tell them apart, especially since each one of them goes by different names. The most popular variety is peppermint, which is also known as piperita mint or Mitcham mint. Each mint has a specific olfactory and gustatory profile; the best way to tell them apart is to bite into a leaf. Peppermint tastes and smells like menthol, spearmint like carvone, and eau de Cologne mint like linalool—their main components, respectively. Menthol is one of the few scents that act on the trigeminal nerves lining the oral cavity, through which we experience the sensation of cold and the pungency of spices. It's the icy mint, the taste of Get 27 liqueur. Carvone, the major constituent of spearmint, is found in toothpaste and chewing gum. Linalool, a molecule with a fresh, flowery scent, is the major component of eau de Cologne mint; it is also found in the essence of bergamot, lavender, coriander, and rosewood. Mint opens a world of marvels.

PEPPERMINT

(Mentha x piperita)

This mint has no shortage of names. The result of natural hybridization between *Mentha aquatica*—which was known as "frog mint" when I was young and has now been discovered to work wonders depolluting soil—and spearmint, or *Mentha spicata*, peppermint was first cultivated in England in the eighteenth century. Naturalist Carl Linnaeus acknowledged *Mentha x piperita* as a species in its own right. Its cultivation and trade as a medicinal plant began in Mitcham, one of the districts of London that gave it its other name. The powerful, fresh taste from the menthol it contains (30 to 60 percent, depending on quality) has made it one of the most coveted plants, earning it a great reputation among herbalists.

Deeming it good for the stomach, French pharmacists matter-of-factly called it English mint because that's where it came from. Perennial and easily propagated, peppermint is cultivated in Essonne,

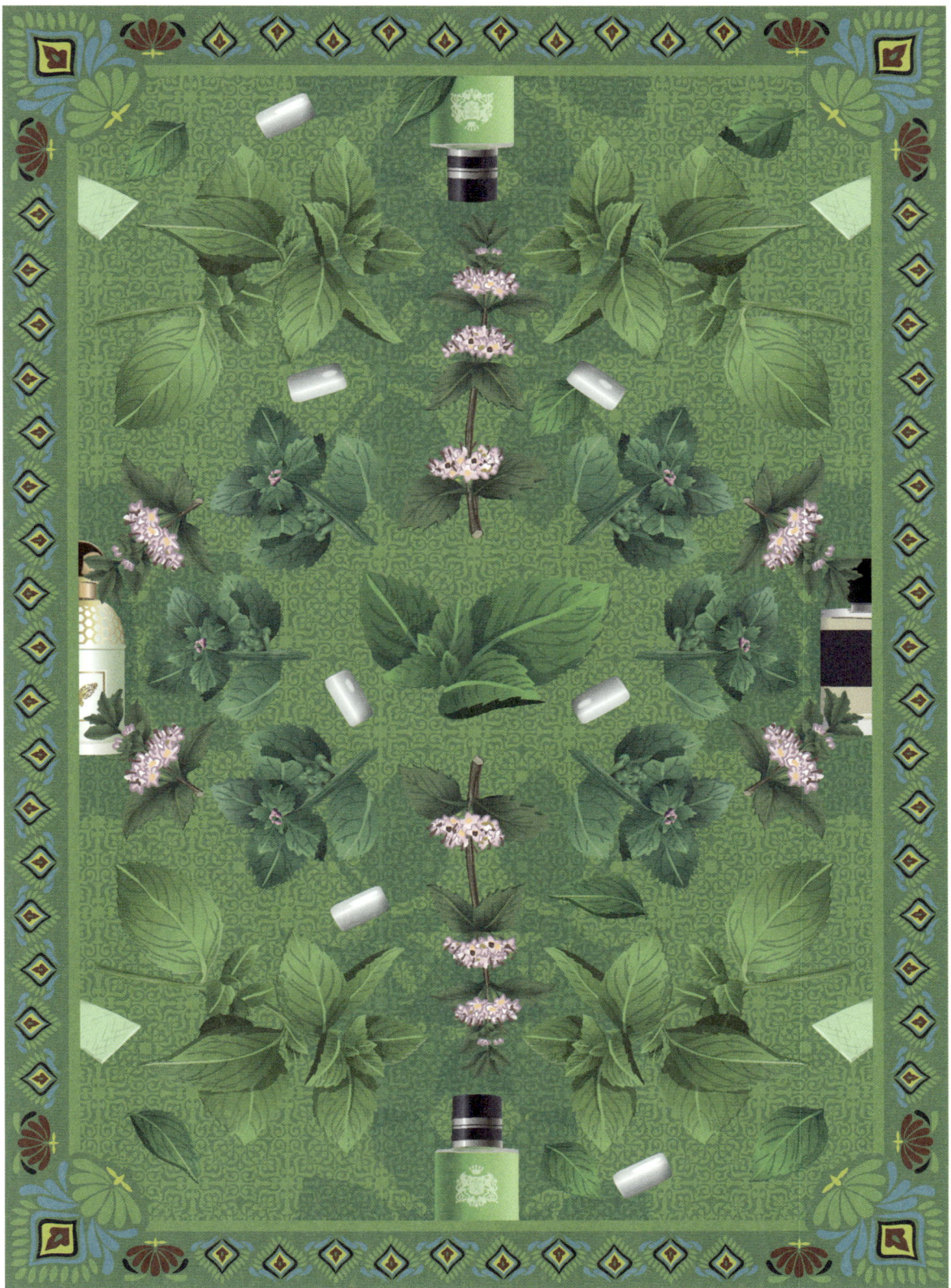

Maine-et-Loire, Drôme, and Alpes-de-Haute-Provence. The essential oil is used mainly for food rather than perfumery. It is found in toothpastes, chewing gums, chocolates (especially the famous After Eight mints), ice creams, beverages such as Ricqlès mint alcohol, and liqueurs. Finally, to give the English and their cuisine due credit, its chopped leaves are mixed with vinegar and sugar to make the sauce that traditionally accompanies leg of lamb. The four main producers of peppermint essential oil today are India, Italy, Argentina, and Australia.

MINT OR SPEARMINT

Mentha spicata

Like its kin, this perennial has many names: garden mint, common mint, and Nanah mint (in Morocco). Easily cultivated, it does well in open ground, preferably with partial shade, or in pots, where it can be picked according to need. There isn't a Chinese person who doesn't have some mint growing in the window, just as there isn't an inhabitant of the Mediterranean without a pot of basil. Its essential oil is obtained by steam distillation; it contains 55 to 65 percent carvone and up to 25 percent limonene (a major constituent of orange oil). Spearmint essen-

tial oil is used in beverages, confectionery, ice cream, chewing gum, and toothpaste. The plant's leaves are key ingredients in the mojito, where they are blended with rum and lime; more recently, they have played a starring role in the Hugo, an Italian cocktail made with elderflower liqueur, prosecco, and sparkling water—a competitor to the famous spritz.

Worldwide production of spearmint essential oil is 2,000 tons, divided between the United States, China, Russia, and India.

FIELD MINT OR AMERICAN MINT, JAPANESE MINT

Mentha arvensis

Like most mints, the *arvensis* species has several names. Field mint is a perennial native to Europe, but it is often confused with a mint native to North America and East Asia known as *Mentha canadensis*, which it resembles. While a high linalool component characterizes field mint, its counterpart is rich in menthol. Linalool and menthol have nothing in common. The latter is what's used for cigarettes, beverages, oral hygiene products, pharmaceuticals, and fragrances. Perfumers who believe they're adding field mint are actually working with Canadian mint, which is often called American mint.

But truth be told, if mint were named after where it comes from today, it would have to be called Indian mint. India is now the world's leading producer, with an annual output of 12,000 tons, having surpassed Japan, the world's leading producer for the first half of the twentieth century.

MENTHE CITRATA

Mentha x piperita var. citrata

The mint Citrata, known as eau de Cologne mint, is also called bergamot mint or orange mint. It is supposed to display the characteristic scent of lemon and orange. In fact, it smells of neither lemon nor orange and has no components of either. Botanists and retailers should enlist the services of perfumers more often so they can come up with a name that facilitates marketing and avoids misleading the public. The name of bergamot mint comes closer since this citrus fruit is composed mainly of linalyl acetate and linalool, which both have a pronounced presence. A new cultivar, "Kirin," was developed by agronomic researchers and put into cultivation in 1988. This plant produces an essential oil containing on average 50 percent linalool and 35 percent linalyl acetate. In India, commercial operations produce between fifty and sixty tons of essential oil a year; mint

Citrata or bergamot mint was introduced in 1959 to the Jammu and Kashmir region, in the Himalayan mountains.

In perfumery, mints are often used in conjunction with each other; peppermint supplies freshness, and spearmint roundness. The best-known mint eau de toilette in France is undoubtedly Green Water, created in 1946 by perfumer Vincent Roubert for Jacques Fath, who wanted to appeal to American clients after the war. The young couturier died prematurely in 1954, but his brand survived and the fragrance was relaunched in 2016. Mint is one of the rare notes specific to eaux de toilette for men or both sexes. Notable examples include Herba Fresca in the Aqua Allegoria Guerlain collection (1999), Menthe Fraîche by James Heeley (2006), Roadster by Cartier (2008), and Géranium Pour Monsieur by Frédéric Malle (2009).

ROSEMARY

Salvia rosmarinus

For Shakespeare, rosemary symbolizes remembrance. Green and full, it seems never to age. Then, all of a sudden, it does so from one day to the next. One morning, we notice that the plant that required nothing—not even water—is in the process of turning brown and gray, drying out, and dying. The best place to get rid of the deceased is not in the compost pile, as its wood is too hard, but in the fireplace. The scent that it emits as it burns is like incense. In any event, rosemary, which can be pruned at will, remains invaluable in the garden for borders and hedges, and as a topiary plant.

A woman strolling by my garden saw the hedge of creeping rosemary whose round, feminine curves hug the flight of steps up to our home. Sitting at the bottom of the stairs, gazing out at the sea, I saw her turn to her companion and say, "That's what I want for our house." I leapt at the opportunity to be of use and observed that it had taken ten years for the rosemary bushes to offer such pleasure to the eye. My remark upset the woman, who seemed impatient by nature, and she promptly changed her mind. The garden she was planning couldn't afford to give so much time to a few rosemary bushes.

Rosemary is a shrub in the *Lamiaceae* or *Labiatae* family. It grows mainly in arid, sunny areas such as garrigues, rocky outcrops, or scrubland, and it requires little care. It's the ideal plant for people who claim they "don't have a green thumb"—a saying that relieves guilt and holds out the prospect of improvement in the future. The ancient Greeks called it *libanostis*, meaning incense-scented shrub, in reference to Lebanon, where incense originated. (Hence "incense tree," in Provençal.) It was also known as *rhops myrinos* (aromatic bush) and *rhus marinus* (sea sumac). The latter approaches the Latin *ros marinus*—a name that derives from the fact that the shrub soaked up the dew of the sea as it flourished on the Mediterranean coast. Jean-Jacques Rousseau wanted to put an end to this onomastic merry-go-round. If botany is to be a science and not just an amateur pursuit, he declared, it must be content with one name for each plant; otherwise, confusion reigns. The most eminent botanist of the day, Carl Linnaeus, said the same thing. Even so, the problem of multiple names persists, giving rise to endless discussions between gardeners, cooks, and sometimes perfumers. Over the two hundred years since that time, science has made advances, and genetics has come to challenge Linnaeus's classifications. But the purpose of this book lies elsewhere.

Rosemary for the perfume industry grows mainly in Spain, Tunisia, and Morocco. Five hundred tons of essential oil are produced each year. It's hard to picture the mountain of twigs the process requires; a high yield of essential oil is just 3 percent. Today, perfumery, aromatherapy, and aromachology all rely on rosemary essential oil. The plant's earliest use along these lines is the stuff of legend. First came so-called Hungary water, or the Queen of Hungary's water, then eau de Cologne, which was invented by the Italian Jean-Marie Farina and marketed in France under the Roger & Gallet brand; the Franco-German alliance was off to a good start. Hungary water, which appeared in the seventeenth century, was a maceration of fresh or dried rosemary in spirits of wine that enjoyed tremendous success. The story behind the name goes as follows. At the end of the fourteenth century, the queen of Hungary was old and sick until a hermit gave her the formula for a miraculous potion for restoring youth and beauty. When she drank the elixir and used it for her toilette, her charms returned. Now a striking young woman, she proposed to the dashing king of Poland, who fell in love with her. This concoction is the magic expedient that Charles Perrault had in mind when he wrote "Sleeping Beauty," not the kiss of some Americanized prince; Walt Disney might not have even known what rosemary is. In the original fairy tale, the princess wakes up only after the spell's appointed duration of one hundred years has elapsed.

For the Egyptians, Greeks, and Romans, rosemary played a role in all special occasions, whether public celebrations, religious festivals, wedding ceremonies, or funerals. Egyptian priests put incense in the sarcophagi of the wealthy, and rosemary twigs in the coffins of those who were less so. Clever Greek students braided it into wreaths that they wore during exams to stimulate their memory. In turn, the Romans, who gave the plant the name it has today, had the custom of adorning bridal couples with rosemary crowns.

Rosemary is featured in Originale Eau de Cologne by Jean-Marie Farina (1709), Eau de Cologne Roger & Gallet (1806), Eau de Cologne impériale by Guerlain (1853), Paco Rabanne Pour Homme by Paco Rabanne (1973), Colonia (1916) and Colonia Assoluta (2003) by Acqua di Parma, and Acqua di Giò Pour Homme by Armani (1996).

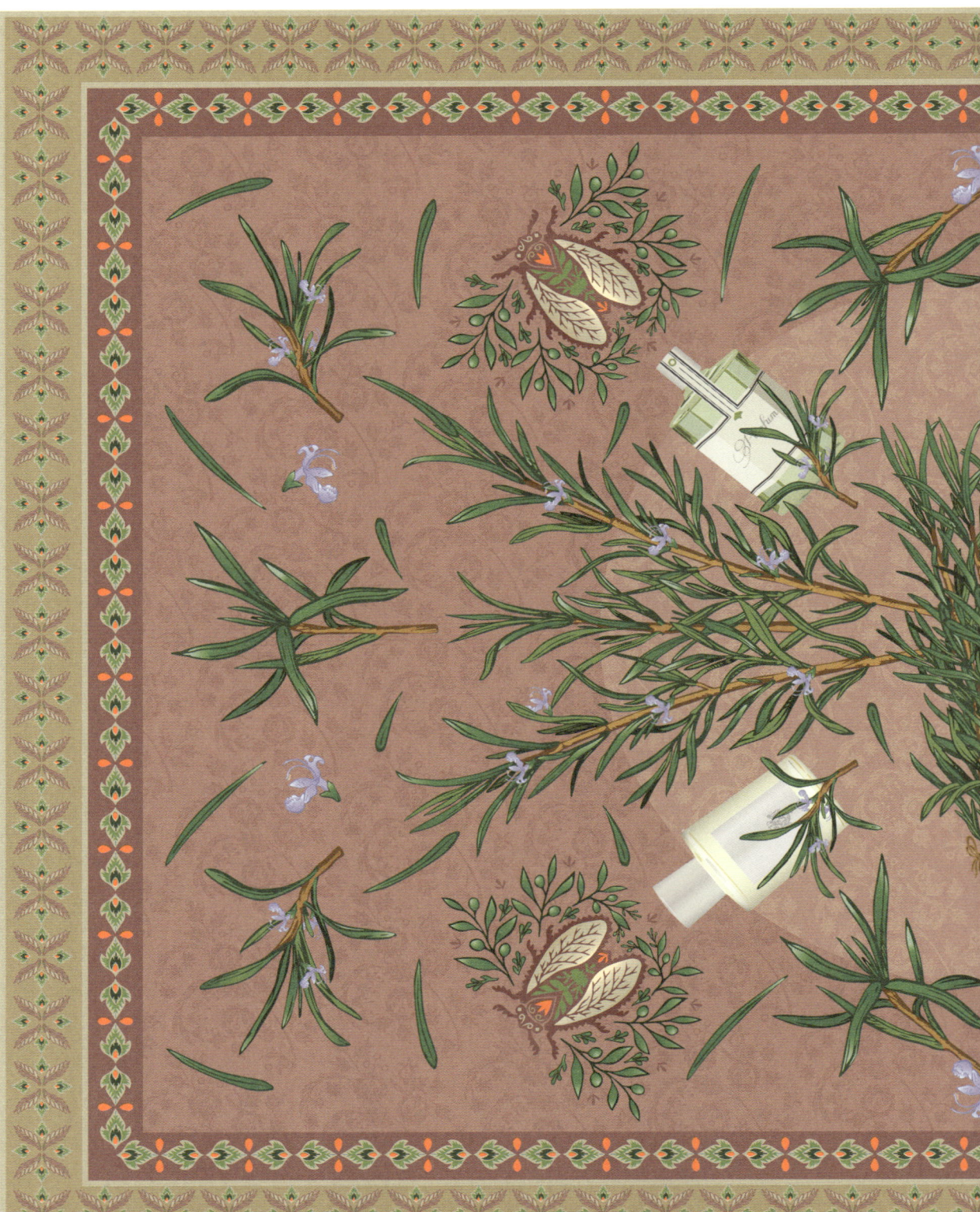

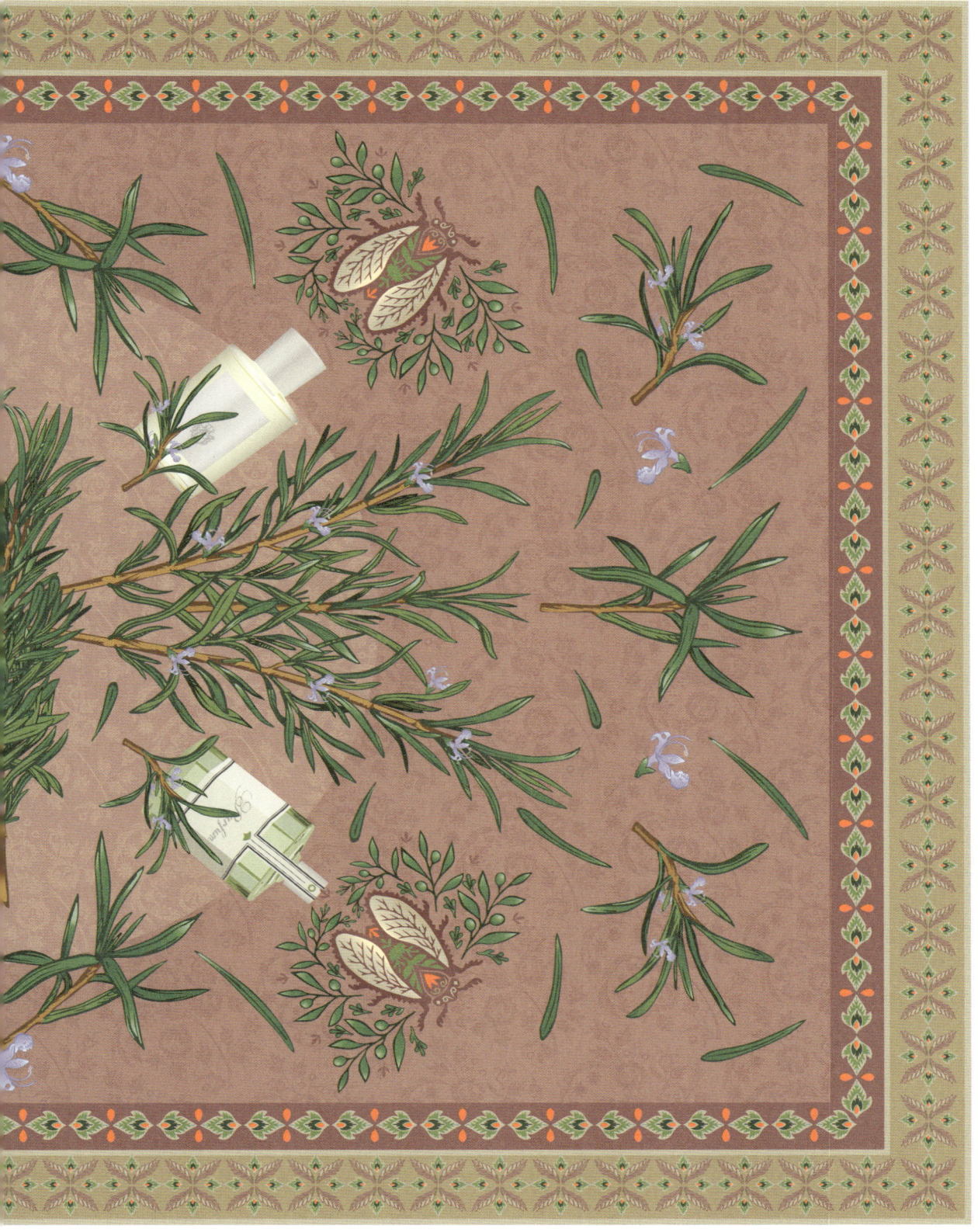

TEA AND MATE

TEA

Camellia sinensis

China is the biggest exporter of tea, surpassing India and Kenya. These three countries together account for 70 percent of global demand. Perhaps surprisingly, Japan comes in only at eighth place. Taking tea, the most widely consumed beverage in the world besides water, became the social ritual of the French bourgeoisie in the late-nineteenth and early-twentieth centuries. For a long time, tea was regarded as an infusion and belonged to the province of herbalists, like lime blossom, verbena, mint, and camomile. It wasn't until the mid-1980s that it attained broad popularity in France, with the opening of houses that offered the best teas in the world, according to the American press.[4] Whereas England still set the tone in Europe, France rediscovered the habit of drinking tea without milk, a practice that both fell in line with Chinese tradition and echoed the spirit of the times. The custom of adding milk originated in India.

Perfumery first acknowledged tea by name in La Route du Thé (1986), when Jean Laporte, the founder of L'Artisan Parfumeur,

received a commission from Barneys, the New York department store. Then came the Thé des Pluies candle (1989), which Olivia Giacobetti created at the request of the Mariage Frères tea house. The scent was conceptual, since no extract existed at the time. It was popularized in 1992 by Eau Parfumée au Thé Vert, the first fragrance put out by the Italian jeweler Bulgari. In the 1990s, accord of tea became a trend, providing a point of contrast with violent fragrances of the materialistic 1980s such as Giorgio by Beverly Hills (1981) and Poison by Christian Dior (1985). It was featured, in unadulterated form, in a whole series of eaux de toilette including CK One by Calvin Klein (1994), Déclaration by Cartier (1998), Green Tea by Elizabeth Arden (1999), Thé Vert by L'Occitane en Provence (1999), Aroma Tonic by Lancôme (1999), and Thé Vert by Roger & Gallet (2000). Indeed, the trend was sufficiently vigorous to prompt raw material manufacturers to produce extracts of different teas. The extracts obtained from leaves served to justify what had been created by illusion. Creating a tea scent does not require tea extract any more than extracts of

---

4. *Newsweek* conferred the distinction on the Mariage Frères tea house for the quality of its darjeelings.

lily of the valley, lilac, or peony are necessary for making perfumes with the scent of lily of the valley, lilac, or peony. But in our own times, when mistrust numbs the imagination, perfumes are like films or novels: they're more convincing if they're based on something real.

MATE

Ilex paraguariensis

Mate is the name of a beverage made from the leaves of the yerba mate plant; along with tea and coffee, it is one of the world's most widely consumed caffeinated beverages. Most commonly drunk in Argentina, Brazil, Bolivia, Chile, Paraguay, and Uruguay, it is also known as Paraguay tea or Jesuit tea. Mate is prepared by infusing roasted, powdered leaves in a gourd; one drinks it through a *bombilla*, a kind of straw that filters the liquid. An evergreen species of holly in the *Aquifoliaceae* family, the tree can reach a height of around twenty meters in the wild. The plant grows at low altitudes in the mountains, most often on the banks of streams. Its proximity to tea is essentially a matter of scent, and its extract in the form of an absolute is nothing new. The first absolutes of mate had a solid consistency and a strong green-black color, which limited their utility. Only when the means were developed to eliminate this color did it become possible to use them widely. Mate absolute has been used in the composition of jasmine-scented bases, and since the 1990s for making tea perfumes.

Perfumes are like films or novels: they're more convincing if they're based on something real.

49

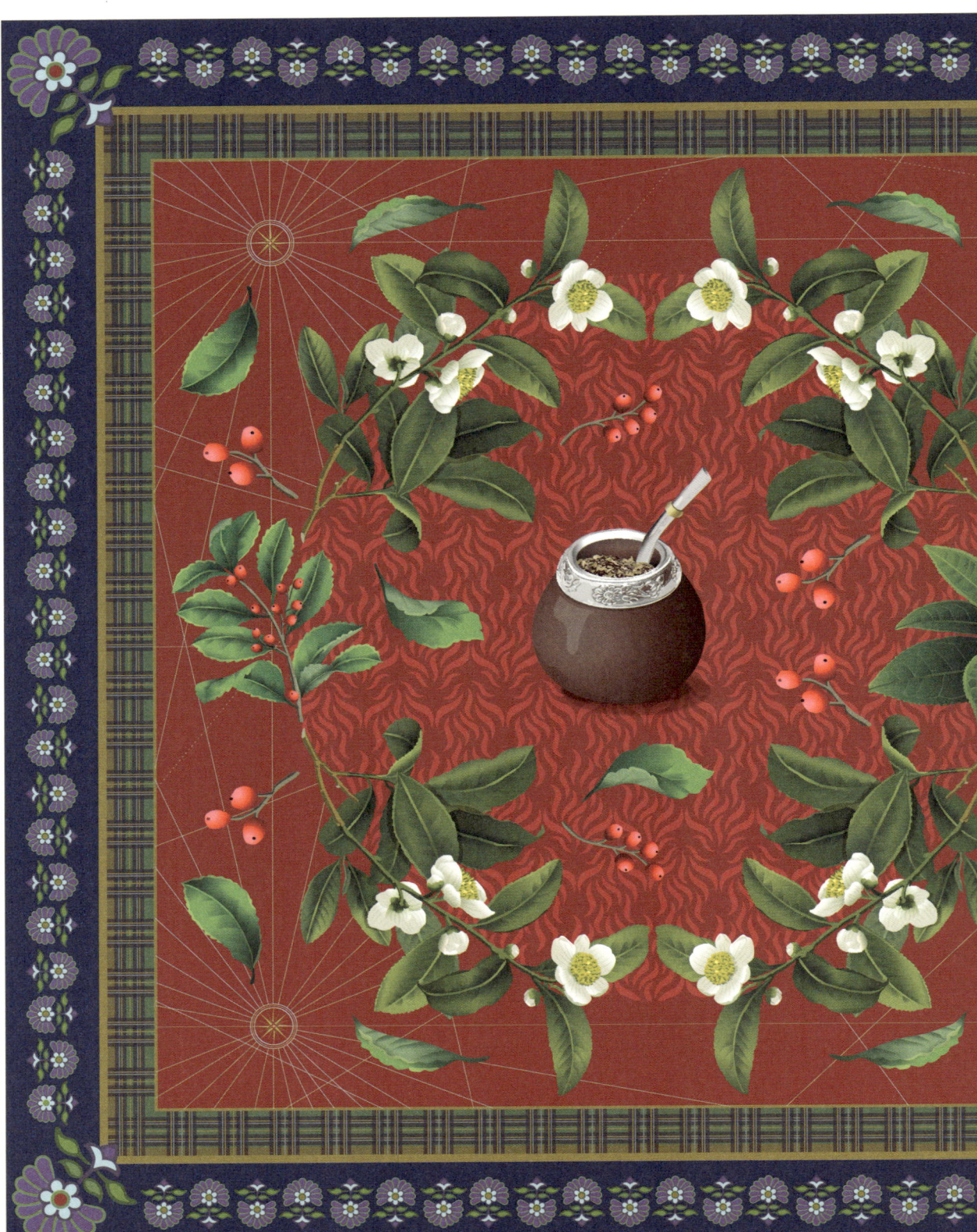

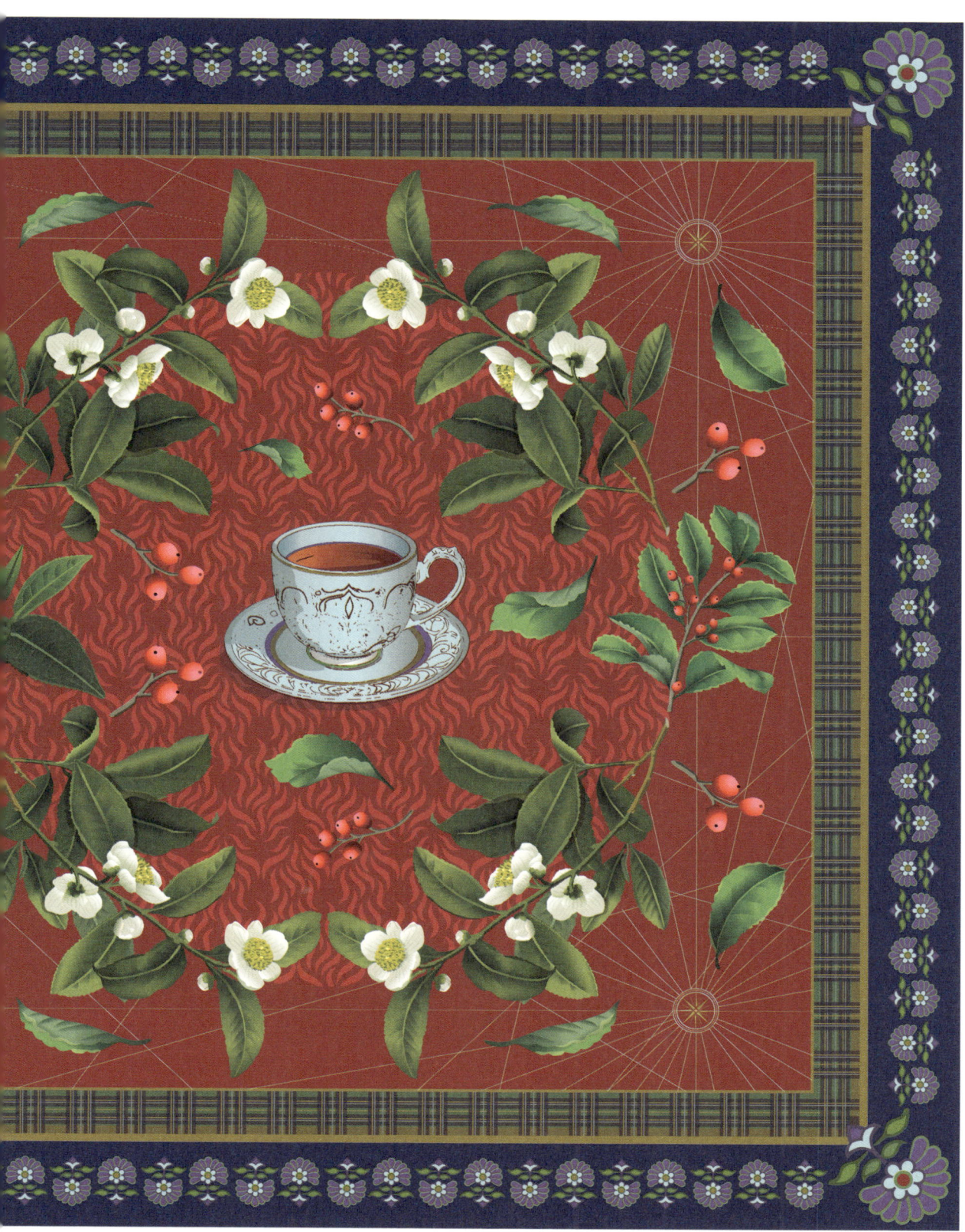

THYME

Thymus vulgaris

Louis XIV never smiled. That seems absurd when one considers the power of the smile revealed by the art of photography. Whether it's Charles Le Brun's portraits of him in his youth, Pierre Mignard's likeness of him posing on horseback, or Hyacinth Rigaud's depiction of him in majestic old age, no portrait exists of the Sun King smiling, which certainly underscores his authority. From an early age, Louis suffered from bad teeth, which made him swallow more than he chewed. By the age of fifty, he had only one tooth left in his upper jaw and his lower

> The Greeks burned it at the altars of the gods and in wealthy homes; they attributed the quality of courage to the plant.

teeth were decayed, so he hid his smile by pursing his lips—which didn't prevent him from having many favorites at court. To ease the discomfort and mask foul breath, he was given thyme, copious amounts of it, as well as cloves to chew on. These two aromatic substances contain phenols, one

thymol, the other eugenol, which have antiseptic properties and soothe aches and pains; some of us will remember the taste and smell of cloves from trips to the dentist. The Sun King's breath may have been fragrant, but chewing high doses of thyme and cloves irritates the palate terribly; perhaps the problem was compounded by his preferred foods: salads and soups. Louis XIV died in 1715 at the age of seventy-eight, after seventy-two years on the throne. There is no evidence that the consumption of thyme or cloves prolongs life.

Our word thyme seems to come from the ancient Egyptian *tham*, the name of a plant used to embalm corpses; it might also derive from the Greek root *thy*, which refers to the exhalation of an odor. The Egyptians used this herb in ointments in order to preserve the dead. The Greeks burned it at the altars of the gods and in wealthy homes; they attributed the quality of courage to the plant. Such symbolism lived on in Europe right up to the time of the Crusades. Damsels would embroider bees fluttering around a sprig of thyme on the scarves they offered to knights headed far away, to so-called barbarian lands. The same bravery was evident in Provence in 1851, when peasant militias fighting against Louis Napoléon Bonaparte's coup d'état made thyme the emblem of their struggle; this sign had been inherited from the revolutionary period, when the Republic was expected to "blossom again."

Botanists consider thyme an under-shrub—a designation that those of a politically correct sensibility might consider derogatory. It just means that the plant has a height of less than forty centimeters. Native to the Mediterranean basin, it grows wild on limestone and clay soils, forming low clumps that are twenty to twenty-five centimeters wide; the plant is ramified and very woody at the base, with small grayish-green leaves covering most of its limbs. Thyme is

It is used in both women's and men's perfumes.

easy to prune into a ball and provides an alternative to boxwood, which has been suffering devastation from invasive Asian moths (*Cydalima perspectalis*) since 2007.

Thyme propagates by division, and it can grow at altitudes of up to 2,000 meters. In spring, tiny white or pinkish-mauve flowers bloom in spikes in the leaf axils, sometimes in the crowns, and bring forth small, round seeds. Over 250 varieties have been cata-

loged, but only *thymus vulgaris* (common thyme), *thymus serpyllum* (wild thyme), and *thymus citriodoro* (lemon thyme) play a role in perfumery. *Thymus vulgaris*, also known as red thyme, is the most commonly used and is often rectified to be sold as white thyme. Its main components are thymol, carvacrol, and geraniol. Steam distillation generally takes place from May to early July. France, Spain, and Morocco are the main producers of essential oils. It is used in both women's and men's perfumes: Essence of thyme can be found in Guerlain's L'Heure Bleue (1912), Loris Azzaro's Azzaro Pour Homme (1978), and Cartier's Must (1981). More recently, wild thyme has been featured in Ocean Rain by Italian bootmaker Mario Valentino (1990). This fragrance was the last creation of the great perfumer Edmond Roudnitska. It launched the wave of perfumes with marine notes that flooded the market in the years that followed; these scents included Armani's Acqua di Giò Pour Homme (1996), a worldwide success that contains rosemary rather than thyme.

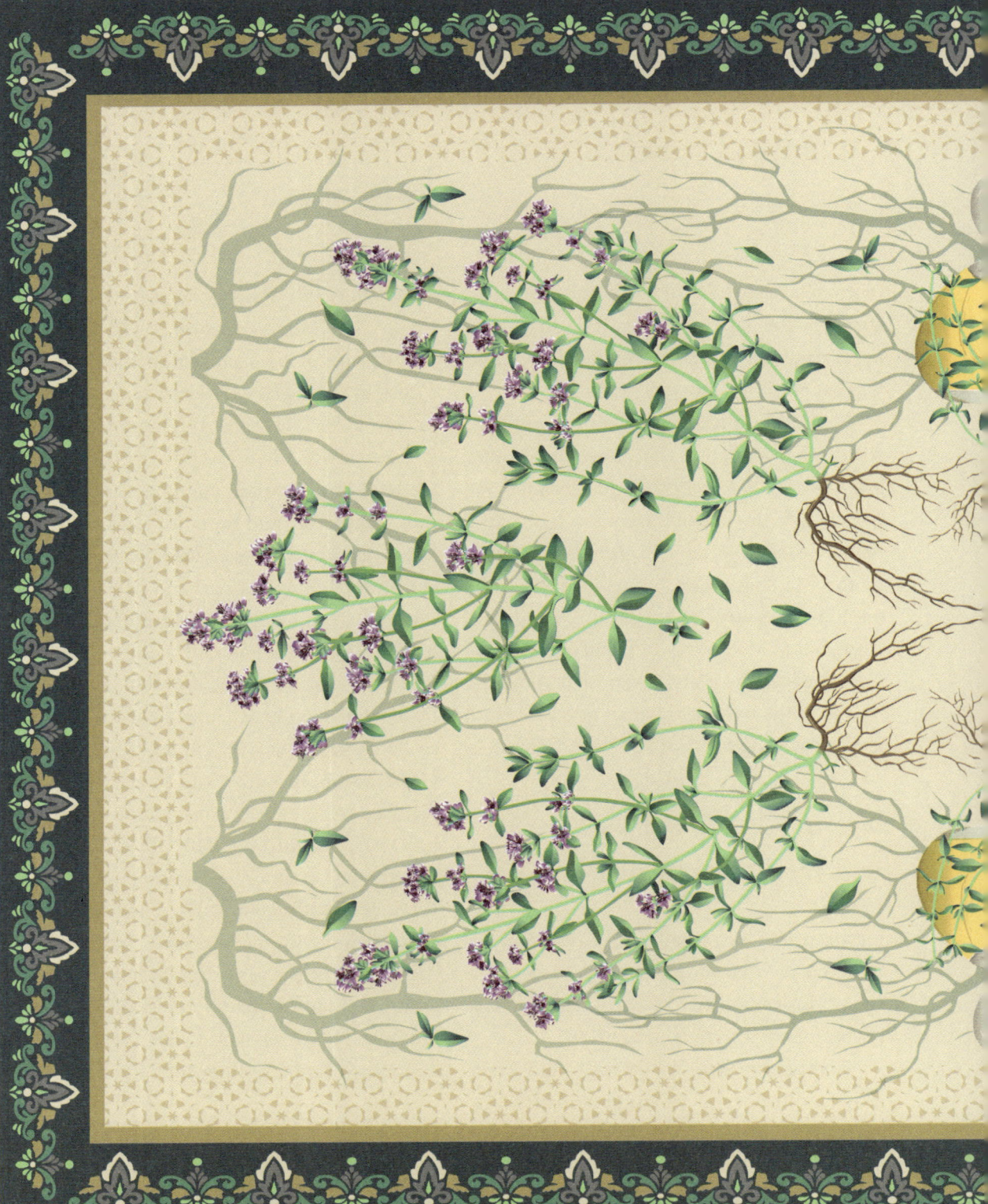

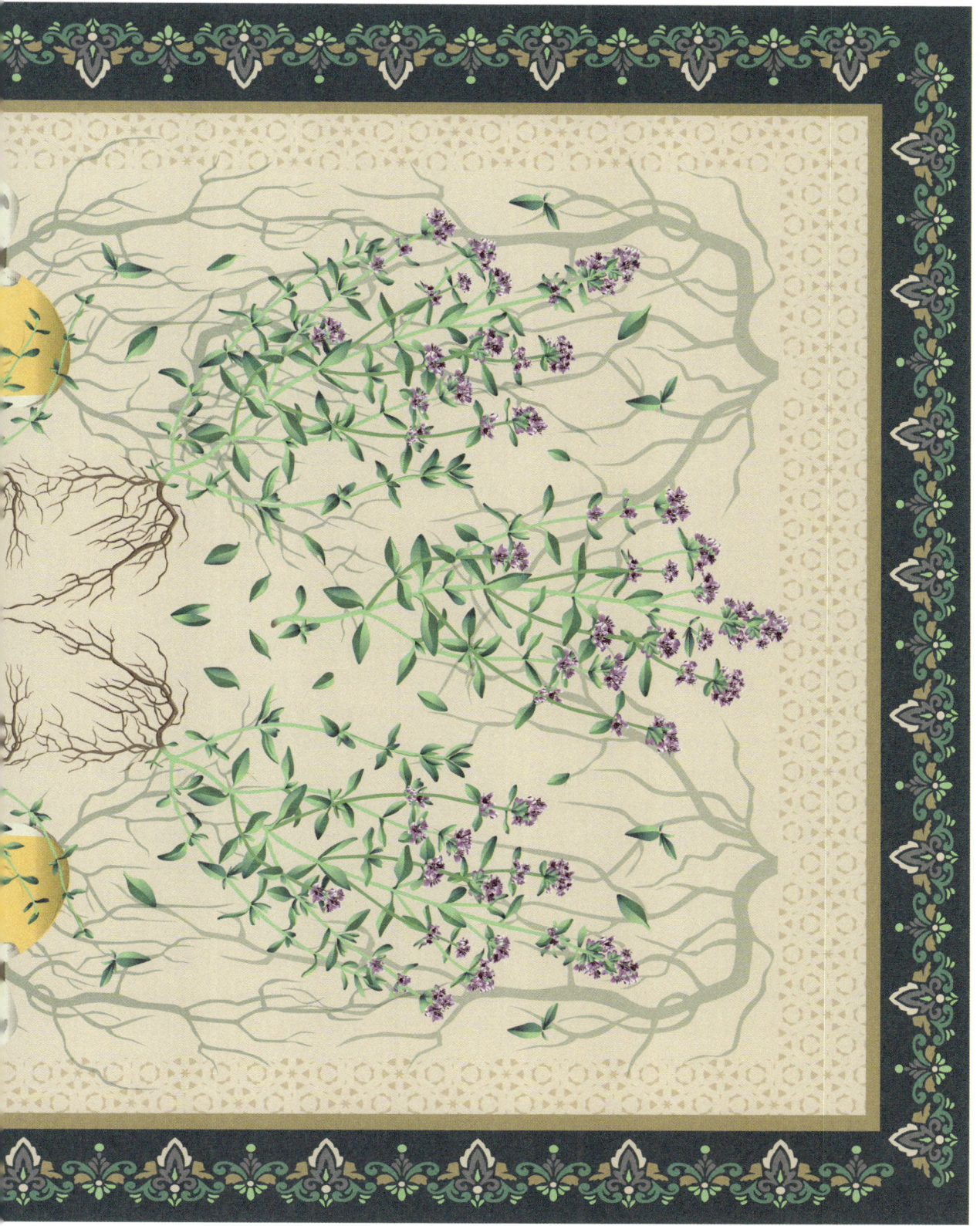

HAY

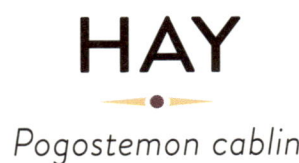

Pogostemon cablin

I'm no botanist. Scents are my domain, and I know plants more by how they smell than by what they're called. After all, perfumes are composed with scents, not names. I have a particular love for the smell of hay. It is redolent of summer, siestas, and amorous play. Monet responded to this sense by painting a series of twenty-five haystacks at different times of the day. As he did so, he couldn't have been insensitive to the powerful, concentrated odor they emitted.

Hay is made from vegetation that is mowed and dried; it is used primarily to feed animals. Unlike other substances used in perfumery—and this is a unique feature—it already represents a composition, since it is a mixture of grasses (*Poaceae*) and legumes (*Fabaceae*). The former include sweet vernal grass, which is laden with coumarin, the molecule that gives the signature scent to hay (and to the tonka bean, or coumarou, from which its name is derived). Hay from La Crau is a favorite among cooks, since its scent brings a whole landscape to the plate. I recall how Pierre Gagnaire, a Michelin-starred chef, paid tribute to Terre by Hermès (2006), a cologne he was fond of wearing. As part of his "Earth" menu, Gagnaire created a dish that incorporated the smell of hay, the odor of the fields, when one lifted the lid of a small cast-iron pot. Made from a wide variety of flora, Crau hay received AOC and PDO status in 1997. Its quality is such that it is now sold in the United Arab Emirates to feed the thoroughbred horses so highly prized there.

I remember, in the summertime, the factories that produced raw materials for perfumery adding lavender to the hay. The bales from the plains of La Crau weighed several dozen kilos—sometimes more—and were tied up with red and white string. This was a sign of their origin and quality, and there was always a numbered certificate, too. The bales then had to be divided up and shredded; with pitchforks, workers filled 2,000-liter stills (the equivalent of twenty bathtubs), each of which held 400 kilos of hay. The yield of essence was low, but since people liked the smell, co-distillations were performed: hay was mixed with cedarwood and also with geraniol, a rose-scented molecule, which produced seductive results. These products remained popular until the end of the twentieth century. Subsequent

Hay is redolent of summer, siestas, and amorous play.

technological advances led to making extract of hay with volatile solvents. Most often, the process is carried out cold to ensure that the scent will be as close as possible to the source material. All distillation produces an artifact—something that doesn't exist in nature. The result was perfect: never has an extract been closer to the original.

In the spring of 2002, one of the songs ringing in my head was "Couchés dans le foin."[5] It was performed by Mireille, a singer who belonged to my parent's generation and had the joyful voice of someone who would never grow up. Her name will mean nothing to my grandchildren or young people today. The refrain goes:

Lying in the hay
With the sun looking on
A little bird singing in the distance
We tell each other secrets
And make great oaths and vows
Our hair's full of straw
We kiss and we wriggle
Ah! that life is sweet, sweet
Lying in the hay with the sun looking on

These charming verses, which may seem childish, inspired L'Eau d'Hiver (2003) by Frédéric Malle, a fragrance that was anything but. In this scent, I sought to express the tender, joyful caress of summer air—which at the same time heralds the turn of seasons. Obviously, a single substance doesn't make a perfume, but hay absolute played a role. It was never found by chemical analysis. The tools were not sensitive enough to understand that the coumarin came from hay extract; to the dismay of those who tried to plagiarize the perfume, the honeyed tenderness of hay was irreplaceable. To imitate is to be inspired, that is, to be creative; copying means the opposite.

5. "Couchés dans le foin" (1930), lyrics by Jean Nohain, music by Mireille.

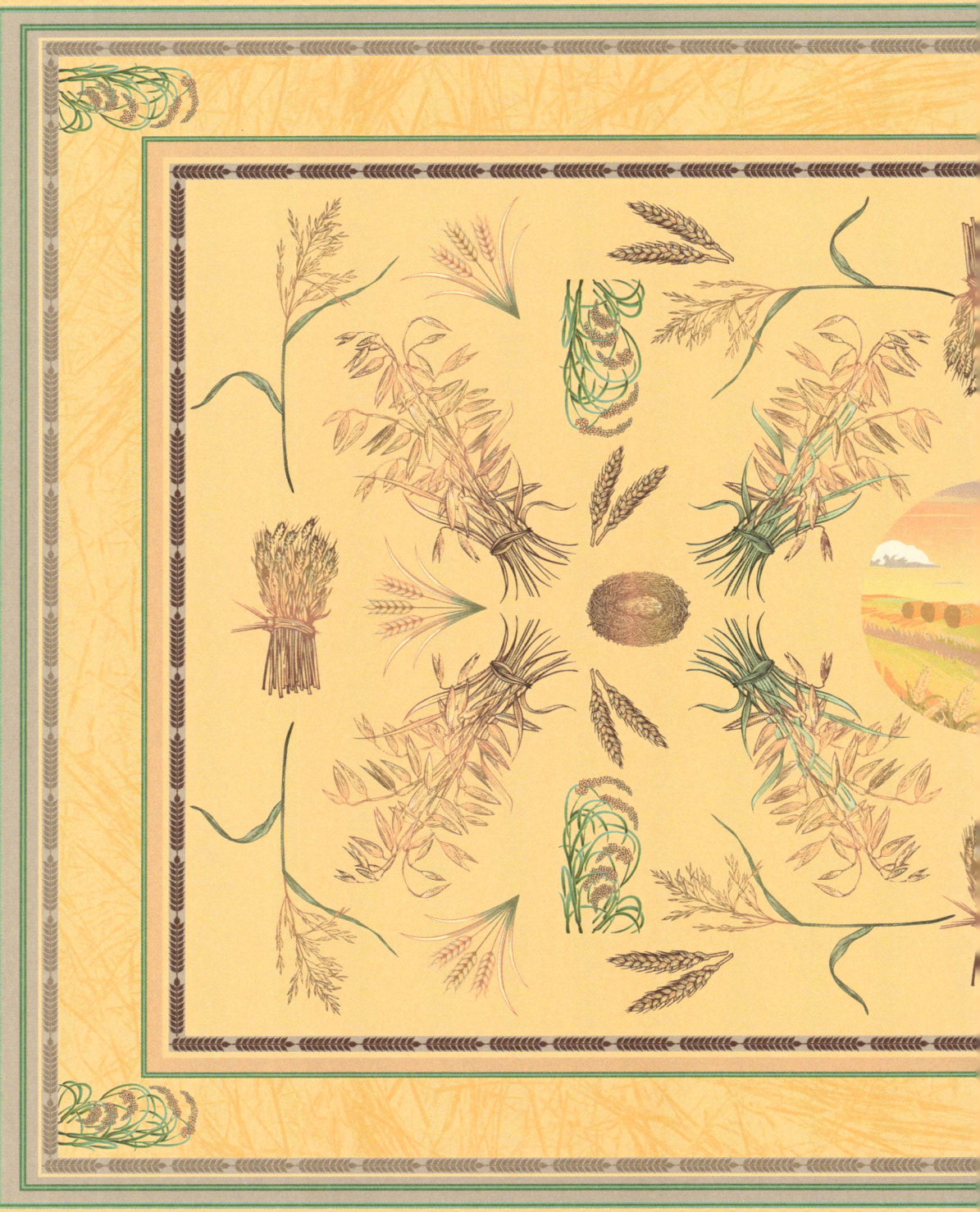

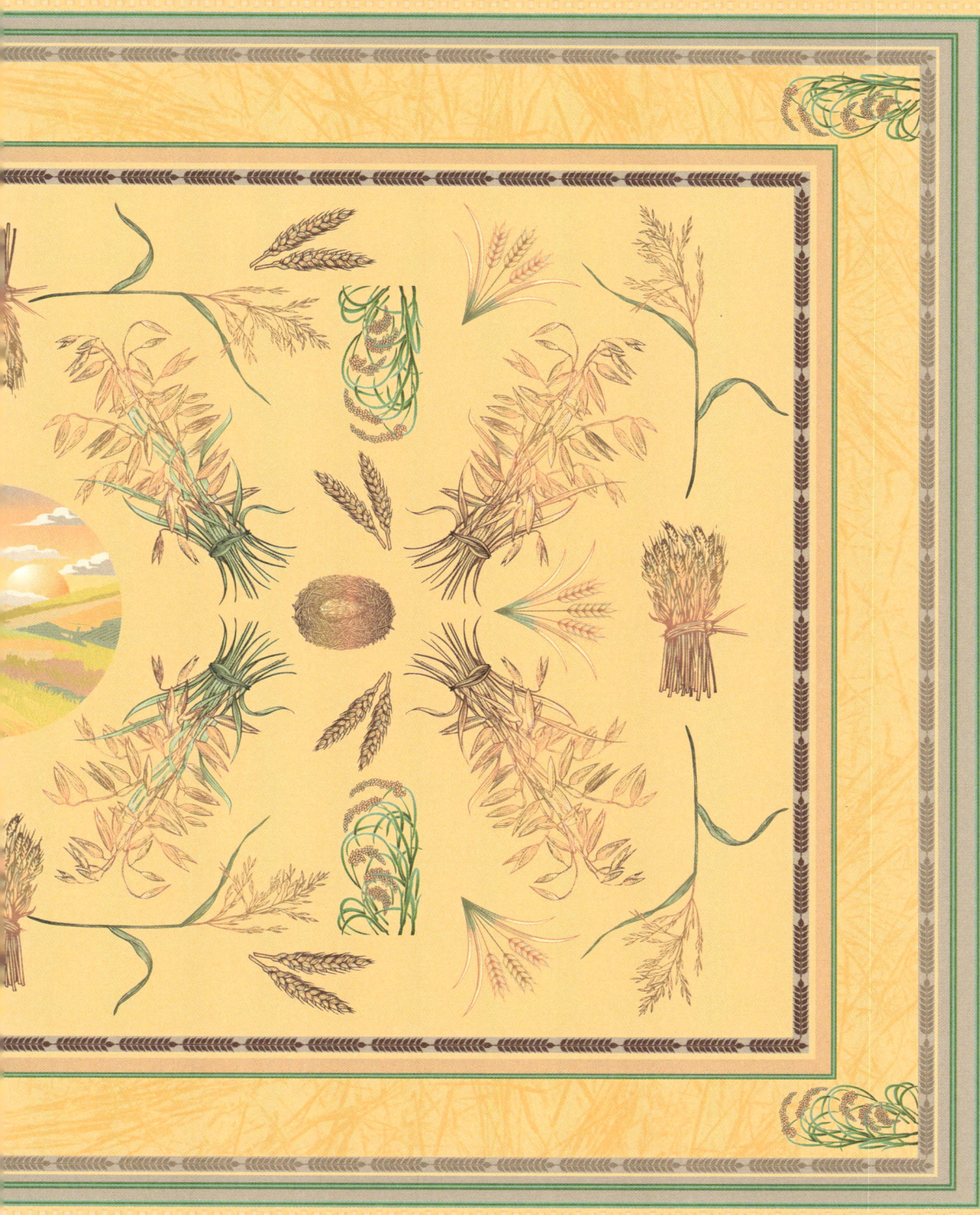

Simple or compound corolla of certain plants,
usually fragrant and brightly colored.
Littré Dictionary

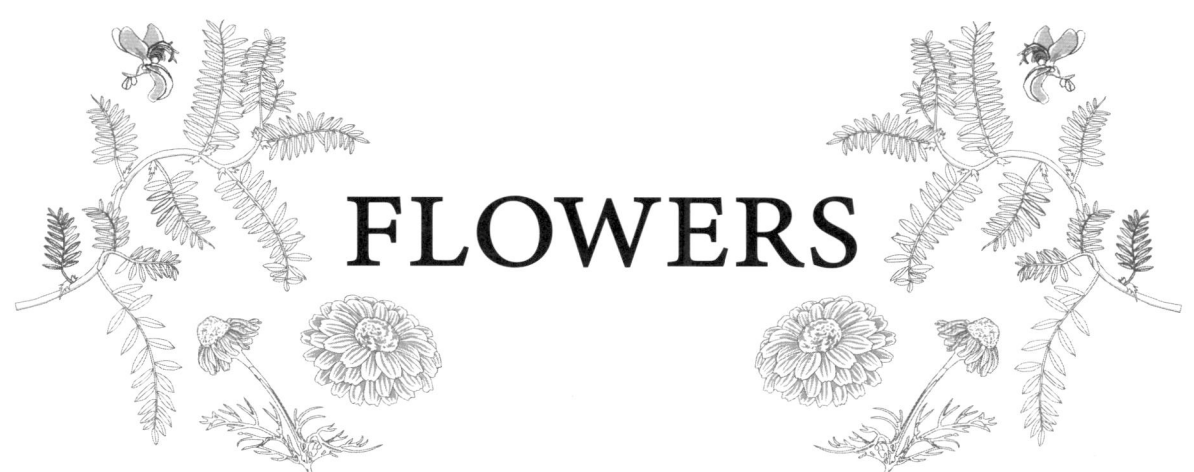

FLOWERS

Very early on, flowers commanded the attention of both women and men. At the prehistoric site of Shanidar in Iraqi Kurdistan, archaeologists discovered the so-called Flower Tomb from fifty thousand years ago. Excavation revealed pollen from hundreds of flowers that once formed a litter to bear the body of someone of high lineage and great dignity. The flowers had not been chosen at random. They all had bright colors and medicinal properties and were perhaps also used for shamanic rituals. So a predilection for flowers reaches far back in time. Flowers form part of our daily lives. Not only do they serve decorative purposes in our gardens and homes; they also provide a source of inspiration to all artistic professions for their colors, shapes, tastes, and smells.

Although all flowers represent sources of creativity and inventiveness, relatively few have been made into an extract. Of the hundred or so raw materials that form the basis of perfumery, only some ten are flower extracts. The main reason is economic. Either the yield is too low, making the cost excessive, or cultivation is too complex, requiring huge areas in order to be profitable. For example, three hectares of rose fields yield four tons of flowers, or 200 million petals. This is only enough to produce one kilogram of essential oil. One gathers just how vain we are to perfume ourselves.

CHAMOMILES

Two main kinds of chamomile are cultivated: Roman chamomile and blue chamomile. Both find use in cosmetics, dermatology, and, to a lesser degree, perfumery.

ROMAN CHAMOMILE

Chamaemelum nobile

If you're unlucky at cards, Roman chamomile flowers may be the answer—not as a cure for gambling, but because the flower is a symbol of prosperity in many countries. According to popular tradition, it is supposed to increase one's luck. People who are superstitious or gullible are advised to arrange dried flower heads in a blue dish (the color of the vessel is not important) and place them in the northwestern corner of the room (this detail is) where the game is to take place. Alternatively, if it's easier, wash your hands in an infusion of flowers to better your odds. I'm not making anything up. I read and I listen. Sometimes I come across extravagant beliefs and practices that give me pause, even though I remain incredulous.

In Latin countries, chamomile is a noble plant. It is valued in cosmetics for its lightening and calming properties. It's also made into tea as a home remedy; indeed, it has proven so popular in France that we often use the word "chamomile" instead of "infusion" to refer to verbena or linden. In Provence and Italy, where amaro is enjoyed daily, it plays a part in the composition of aperitifs. Bitterness delights me because it's an intelligent taste and smell. Unlike sweetness, it poses questions. It features prominently in vermouth and Apérol, which is combined with prosecco to make the spritz, a bubbly and cheerful cocktail that has become the world's most widely consumed alcoholic beverage.

In perfumery, essential oil from Roman chamomile is one of the "trace" products that chemical analysis cannot detect because it is used in low doses; still, it produces seductive and lasting effects in floral fragrances, especially those with notes of rose or clary sage. Roman chamomile can be found in Équipage by Hermès, a scent composed by Guy Robert in 1970 that is a variant of the previous year's Monsieur de Rochas, and in Van Cleef & Arpels Pour Homme, which was created in 1978. As fashions have changed, chamomile has been replaced by magnolia leaf and flower essences, whose scents are similar.

Pliny the Elder said that chamomile smelled like apples, and I wouldn't deny it. Its major components, isoamyl angelate, tiglate, and isobutyrate, have the scent of overripe fruit. A perennial herb in the *Asteraceae* family, the plant is native to the Atlantic seaboard of Europe (Portugal, Spain, France, the United Kingdom, Ireland) and North Africa (Morocco, Algeria). It is scarcely Roman, then. The best-quality French chamomile is grown in the Chemilloise region; it takes the name "Anjou chamomile" from its abundance in this province. A plant with strong, shaggy roots, it thrives in ordinary, sandy, well-drained soils and is harvested from June to July, when it begins to bloom. Flowers bunch together in solitary heads at the end of branches, which measure twenty to thirty centimeters in length.

BLUE CHAMOMILLE

Matricaria chamomilla

Chamomile matricaria is an annual herbaceous member of the *Asteraceae* family with a single stem. Twenty to sixty centimeters high, it blooms from May to November. Compared to those of Roman chamomile, its flowers are tiny; they can be recognized by their typical scent. The plant has no known birthplace and can be found everywhere—in Europe, China, North Africa, India, and North and South America. But its preferred home is Bulgaria, Romania, and above all Hungary, where it has been grown commer-

cially since antiquity. Pliny the Elder already described its medicinal uses. Blue chamomile is also called German chamomile and wild chamomile. The name comes from the high percentage of azulene in the plant's composition, up to 12 percent; the word *azul* means "blue" in Spanish. Azulene is known as a means of soothing skin, combating inflammation, and decongesting the eyes. It has long been added to laundry detergent to lend blue highlights to sheets, which brings out whiteness. The human need for enhanced reality is nothing new.

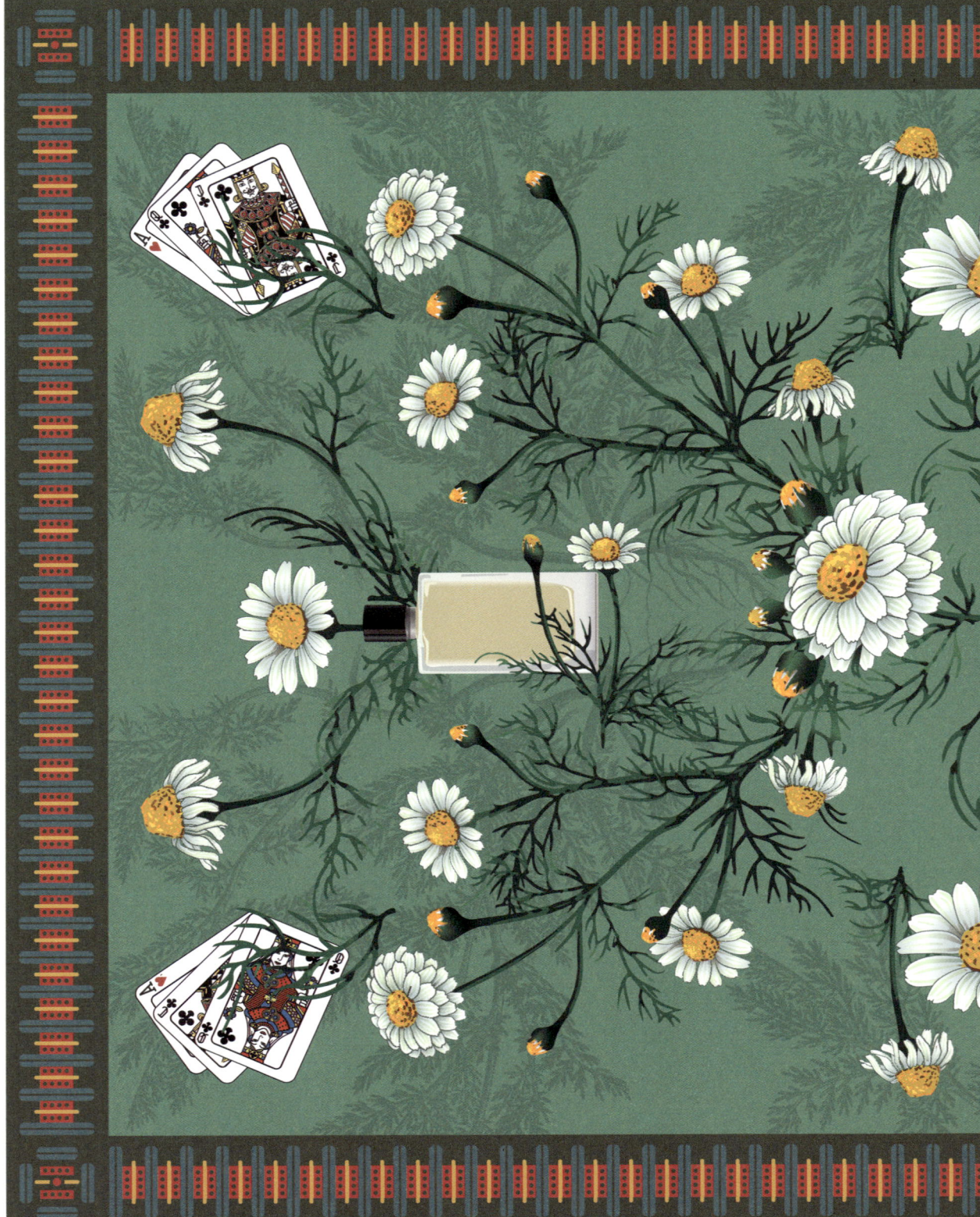

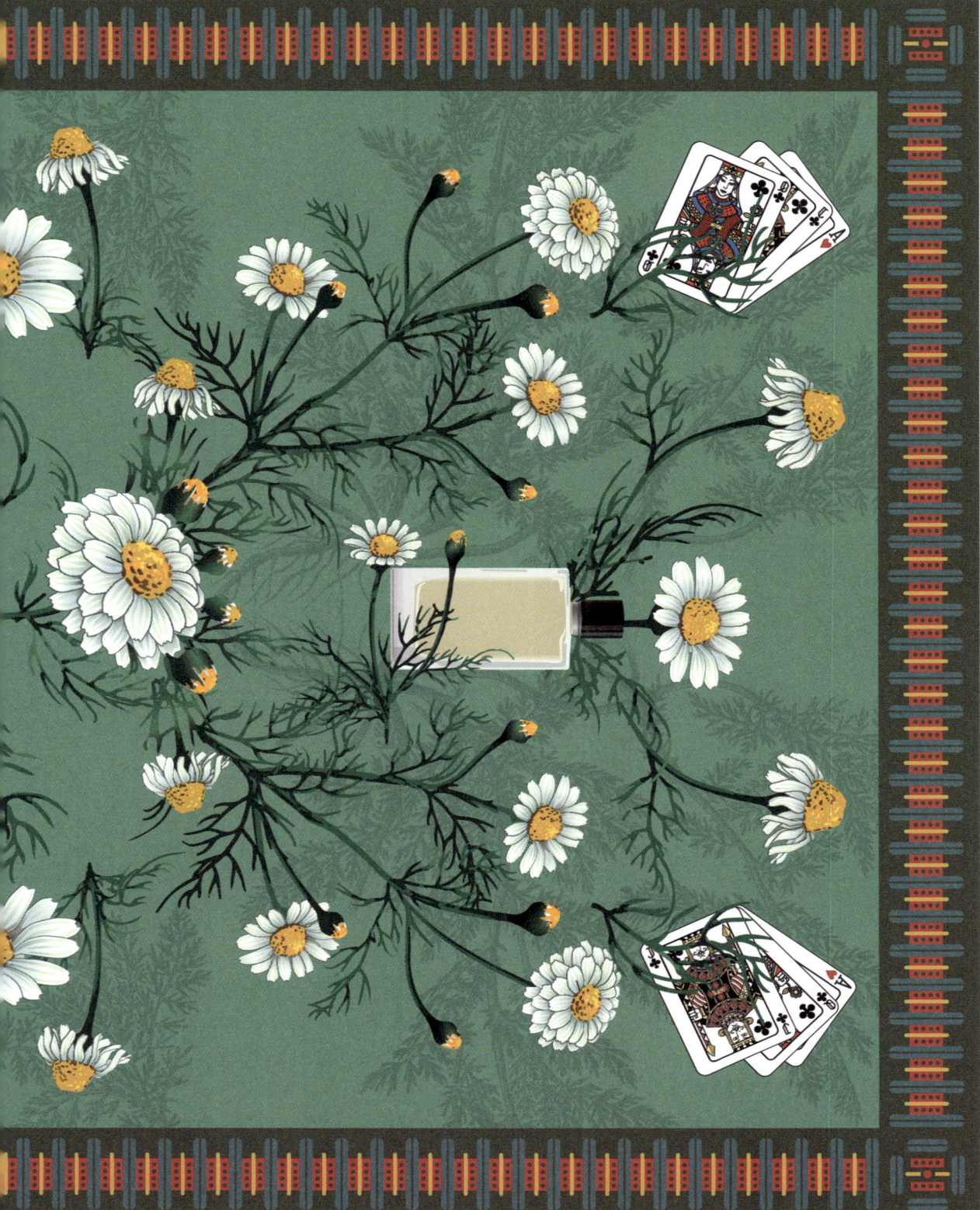

SWEET ACACIA OR CASSIE

Vachellia farnesina

Un jardin sur le Nil, composed in 2004 for Hermès, was my introduction to Egypt and the Nile, that long, quiet river whose floods make the thin strip of land on either side so fertile. Our first grandson was in sixth grade and studying Egypt at school. The trip was his tenth birthday present. We sailed slowly up from Cairo to Aswan; the black water flowed smoothly under the hull. The light was bright and the wind was dry; the ship's diesel fumes masked what smells there were. Tamarisk trees provided shelter to white egrets; from a distance, they reminded me of the handkerchief trees I'd discovered in the gardens bordering the great Italian lakes. Acacias, or cassia trees, dipped their spindly arms covered with sparse flowers along the banks. I could only marvel, since they bloom from November to December, and it was mid-April. I brought up the matter with the captain, who—per the dictates of modern tourism—spoke perfect English. He confirmed my observation and said I should come back in December, when the banks of the river turn golden yellow; this thorny tree provides a fence surrounding many properties. He added that if we did return, we could see the first threshing of the trees and the harvesting of flowers; the second threshing would occur in mid-January, with intoxicating smells reaching as far as the boat.

In winter, there's nothing to be done in the fields; extra income is welcome to workers. By his account, close to one hundred tons of flowers are gathered. The figure astonished me. When you consider that a flower is as light as a feather, it's hard to picture a massive pyramid of them. During a pleasant stop on our journey, I found a tree that was partially in bloom. The flowers, which displayed the color of a young chick, looked like clusters of tiny pompoms. They were very soft to the touch and had a scent of matching sweetness. I was familiar with acacia extract; this absolute has a fragrance similar to that of mimosa, but with an added, epicurean note of red fruit, raspberry. Had we come in December, it's possible that we would have encountered a very different garden on the Nile.

Vachellia farnesiana, or *Acacia farnesiana*, is named after Cardinal Alexander Farnese, who introduced it to Rome's first private botanical garden in the late-sixteenth century. A thorny shrub two to four meters high that originated in Jamaica, it has deciduous leaves and tufts of yellow flowers. The foliage is much appreciated by giraffes. Scientists have studied the coevolution that occurred between this plant and its animal assailant. To protect itself from the giraffe's appetites, acacia developed spines; in turn, the giraffe's

SWEET ACACIA OR CASSIE |

tongue lengthened. Then, the acacia made its leaves fold up; the giraffe's tongue became rough, to pry them open. That's where things stand. I have no idea what the next chapter in this "bad romance" will look like.

Acacia was introduced to France a century after Farnese. Along the Mediterranean coast, from Nice to Marseille, it became the Provençal shrub par excellence—a status now enjoyed by oleander. In days gone by, every inn, auberge, and hostelry had its own tree, in plain view and facing due south, as a matter of course. According to legend, "the beautiful women of the people in Marseille picked the yellow flower to perfume their lips and tresses when they passed; they went home happily, *la Cacio à bouco.*"[6] This phrase refers to the scent of a kiss. Perfumers took up the idea of perfuming lipsticks with acacia. Marketing used to be a matter of intuition and timing.

Two varieties of acacia were grown in Grasse before the plant came to be cultivated on the banks of the Nile: sweet acacia, which flowers from May to November, and acacia caven, which starts in September and finishes at the end of January. The two kinds of tree were a godsend for Grasse's industrialists because they prolonged the period during which floral extractions could be made. Production began in May with roses and oranges, and continued in June with daffodils and narcissus; then came broom, lavender, jasmine, and tuberose in summer. The famously icy winter of 1956 proved fatal to the acacia, which is sensitive to cold. Since it requires good soil and a great deal of care—and because its absolute did not find sufficiently broad use—it disappeared from the Provençal landscape.

There's an important link between Chanel No. 5 and acacia. Perfumer Ernest Beaux, the creator of the scent, used high doses of aldehydes, chemical components that increase olfactory volume. Mademoiselle Chanel had asked for a powerful, abstract fragrance. The aldehydes fulfilled her wishes, asserting themselves like trumpets but masking the play of the flowers. To bring back the harmony, Beaux added cassie absolute, which muted the sound and allowed the beauty of the flowers to ring forth.

Some fragrances are unaffected by time, fashions, and trends; they're called classics. Chanel No. 5 turned 100 in 2021. Thanks to this perfume, acacia shows no sign of disappearing.

---

6. Louis Marret, *Les Fleurs de la Côte d'Azur* (Edition P. Chevalier, 1925).

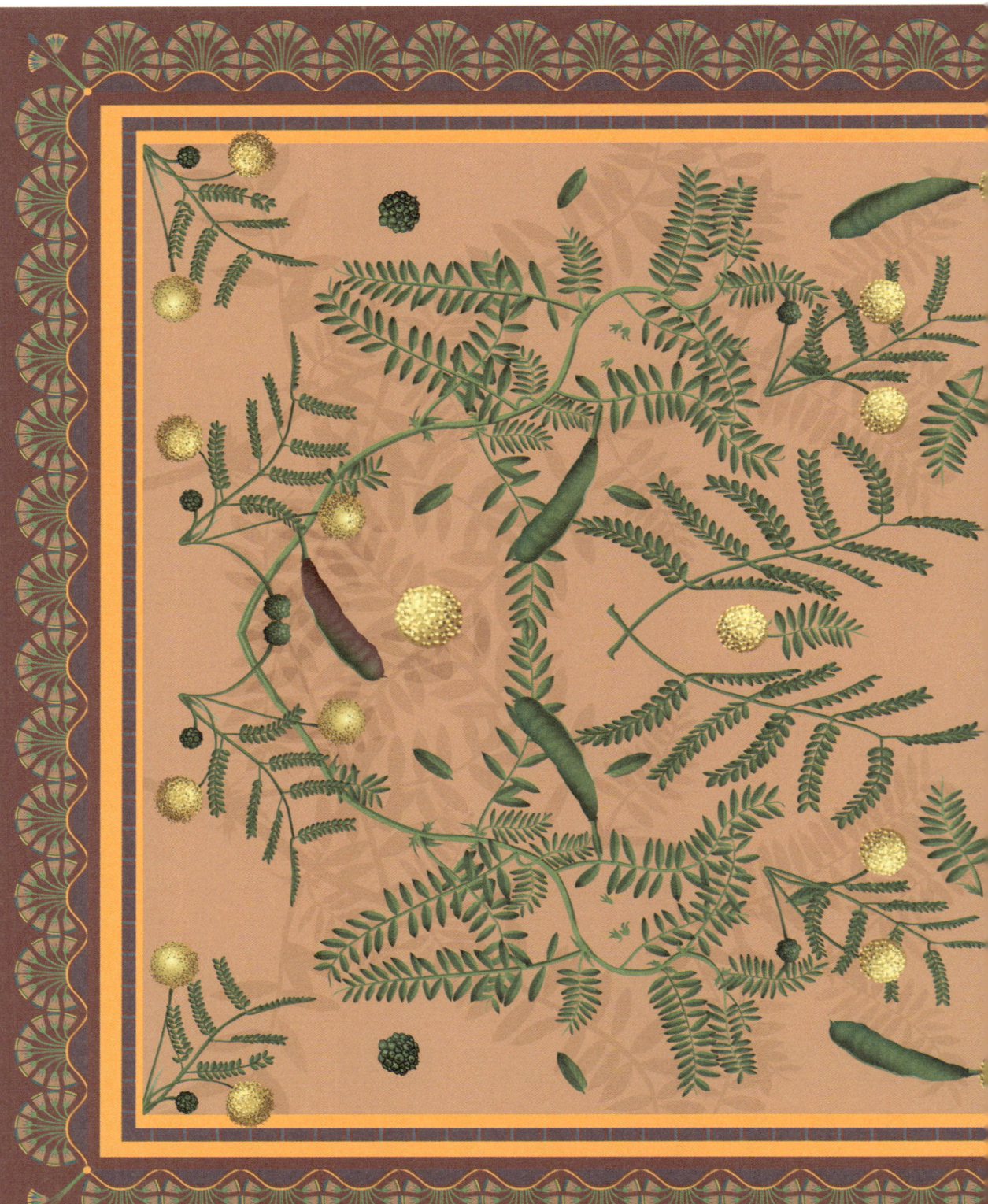

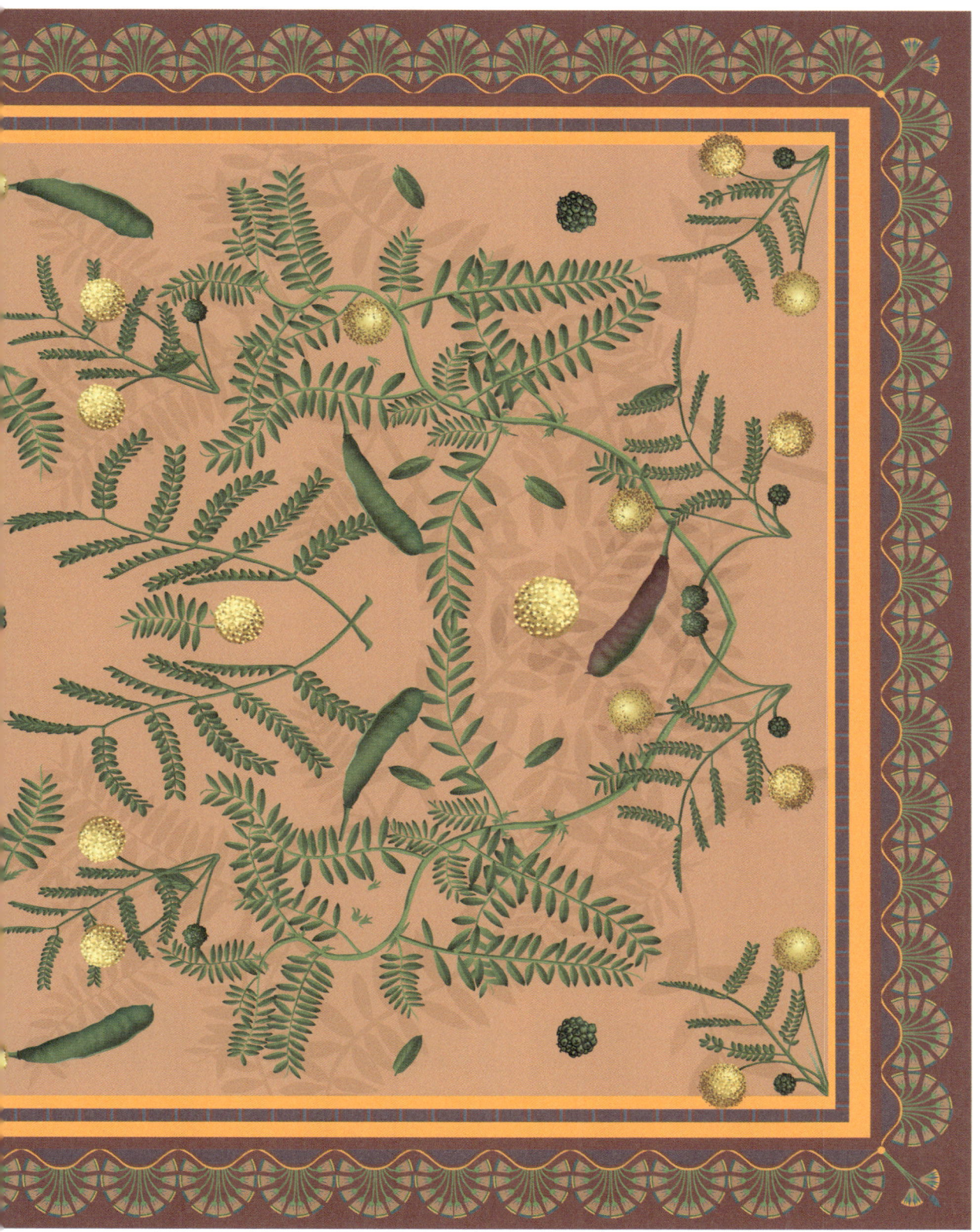

BROOM

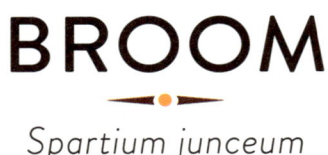

Spartium junceum

The child bent down, closed her hand at the base of a rush-like stem, and pulled it up briskly. She opened her palm and smiled: she was holding something golden yellow with petals that looked like butterflies and smelled of honey and sunshine.

Broom is a bushy shrub in the *Fabaceae* family that can reach two meters in height. It looks healthy for most of the year; in autumn, it turns brown, with green veins. Because of its shape and the color of its bright flowers, it's easy to spot on the plains along the A6 freeway between Aix-en-Provence and Cannes, as well as at the foot of Mont Sainte-Victoire and on uncultivated hills. This hardy plant can withstand severe drought, cold, and even fire. Long associated with witches in the popular imagination, in real life it was used by workers for clearing roads. It also served to make the

> The perfumer recognizes scents that speak or are spoken of, such as jasmine, rose, and tuberose. Others say nothing.

bundles covering the upper part of the millstones used in charcoal production. Finally, its stems have been fashioned into ropes, fishing nets, shoes, and baskets.

Spring harvesting in the Grasse hills lasted for only a few weeks, and broom extract was never made regularly. The local product came to be replaced by extract from Italy, which was more affordable. Poor broom never found its audience—or, more precisely, its perfumer. Even though a few kilograms of extract are still made every year in honor of the formulas for old fragrances, the story makes me think of Aesop's fable, "The Fox and the Leopard." Here, two animals engage in a dispute over who is more beautiful. The leopard boasts of the variegated spectacle presented by its coat. In turn, the fox declares: "How much more beautiful I am than you! I am varied not in body, but in mind." Perhaps broom lacks such cleverness. The perfumer recognizes scents that speak or are spoken of, such as jasmine, rose, and tuberose. Others say nothing; they include broom, longose, lotus, and many others. These scents are beautiful but mute. Their extracts are used to adorn and embellish; they form an ornamental border and a frame for what is being said.

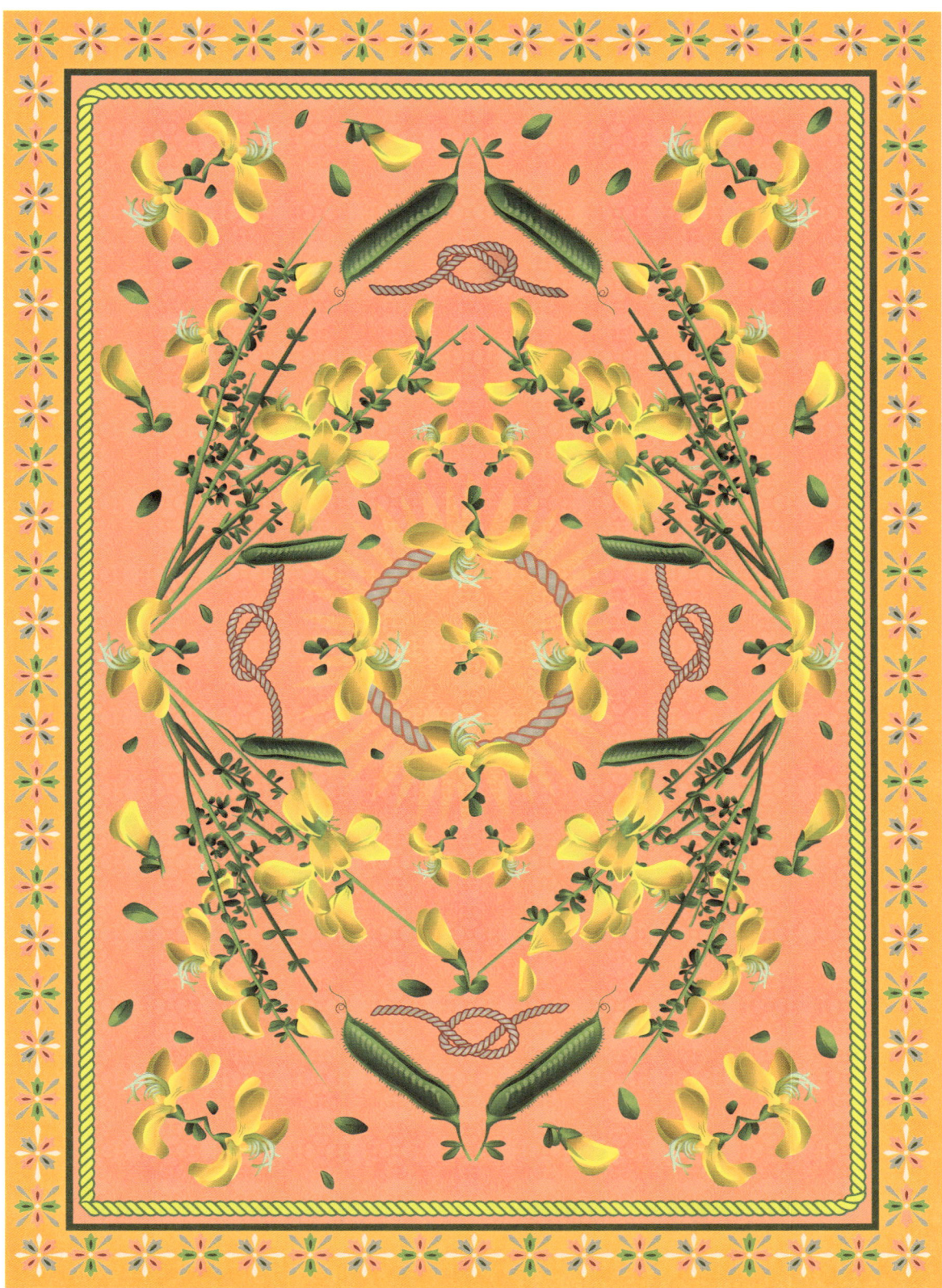

IMMORTELLE

Helichrysum stoechas

Some plants experience belated love that then grows with the years. For them, as for people, hope springs eternal. This is the case for the immortelle or Mediterranean strawflower, which comes in *Augustifolium* and *Italicum* varieties. The name—"everlasting"—heralds glory. Ever since a cosmetics company in Provence placed it at center stage, it has been a favorite with perfumers. In 2022, immortelle received first honors at the International Museum of Perfumery in Grasse. The Respirer l'Art exhibition[7] presented a pollen-yellow, Louis XV armchair quilted with immortelle flower buds atop a lavender carpet: the Platonic ideal of a seat for academic dignitaries. The creations of Peter de Cupere, the Belgian artist who made the piece, highlight the importance of scent in the beauty of objects. Fragrances belong to the realm of art.

The botanical name *helichrysum* comes from the Greek *helios*, meaning "sun," and *chrysos*, meaning "gold" or "golden," the color of its flower buds. Immortelle is a perennial plant twenty to sixty centimeters high, with white, cottony stems. Its leaves, which are rolled at the edges, are green on top and downy white underneath. Small, tightly packed yellow flowers form inflorescences. The flower likes dry, rocky, and even sandy soils; in French, it's sometimes called "dune immortelle." It grows all around the Mediterranean: in France, Italy, Slovenia, Croatia, Bosnia, Montenegro, Albania, Greece, Tunisia, Algeria, Morocco, Spain, and Portugal. Corsica accounts for an estimated 35 percent of world production. Immortelle enjoys protected status in the PACA region, as well. The Haute-Corse and Corse du Sud departments regulate the wild harvest of species that are used in the so-called PPAM (perfume, aromatic, and medicinal plants) sphere; such measures demonstrate the plants' economic significance. Immortelle is gathered from May to early July. Traditionally, the work was done by hand with a pruning hook. Today, it's a machine—like the reaper used for lavandin on the Valensole plateau at the same time of year. Immense golden-yellow carpets give way to the mauve-blue of lavandin. The flowering tips are steam-distilled to obtain essential oil. Volatile solvent extraction produces absolute. Both flowers and extracts give off a spicy scent reminiscent of curry, coffee, molasses, and brown sugar. The reason for this is the fact

7. Respirer l'Art, M.I.P de Grasse, May 2022 to March 2023.

that immortelle contains 35 percent neryl acetate (a fruity scent, like rosé wine), but, even more, 15 percent gamma- and alpha-curcumene. As the name suggests, curcumin smells of turmeric, a major component of curry—which is nothing other than a British appropriation of India's famous garam masala. In this context, it occurs to me that the Roman *garum* condiment, which was made from fermented fish, has a recipe like that of Vietnamese *nuoc mam*. Who did the copying? But I digress.

In the 1970s, I worked for Antoine Chiris, an important firm in Grasse that is now defunct. Exclusive supplier to Coty perfumes, the company produced an absolute of immortelle, as well as a tonka absolute that was used to flavor Amsterdamer. A well-known tobacco at the time, Amsterdamer could be recognized by its packaging: a navy-blue pouch showing a sailor sporting a cap and smoking a pipe. "It smells good, it's Amsterdamer," went the slogan. The scent had a generational significance, and therefore a sociological one, too. Like Gauloises or Gitanes cigarettes, it evoked what was old and established, especially the sturdy working class. Smells are like fashions. They come and go—and sometimes they come back in new attire.

Both as an essential oil and as an absolute, immortelle connotes Indian cuisine. At a low dose, it reveals other, fruity notes such as rose, rockrose, plum, and fig. Immortelle can be found in Femme by Rochas

Both flowers and extracts give off a spicy scent reminiscent of curry, coffee, molasses, and brown sugar.

(1944), Magie Noire by Lancôme (1978), Pour Homme by Van Cleef & Arpels (1978), Sables by Annick Goutal (1985), Eternity for Men by Calvin Klein (1989), Cuir Beluga by Guerlain (2005), Immortelle Corse by Parfums d'Empire (2019), and Ambra by Le Couvent Maison de Parfum (2021).

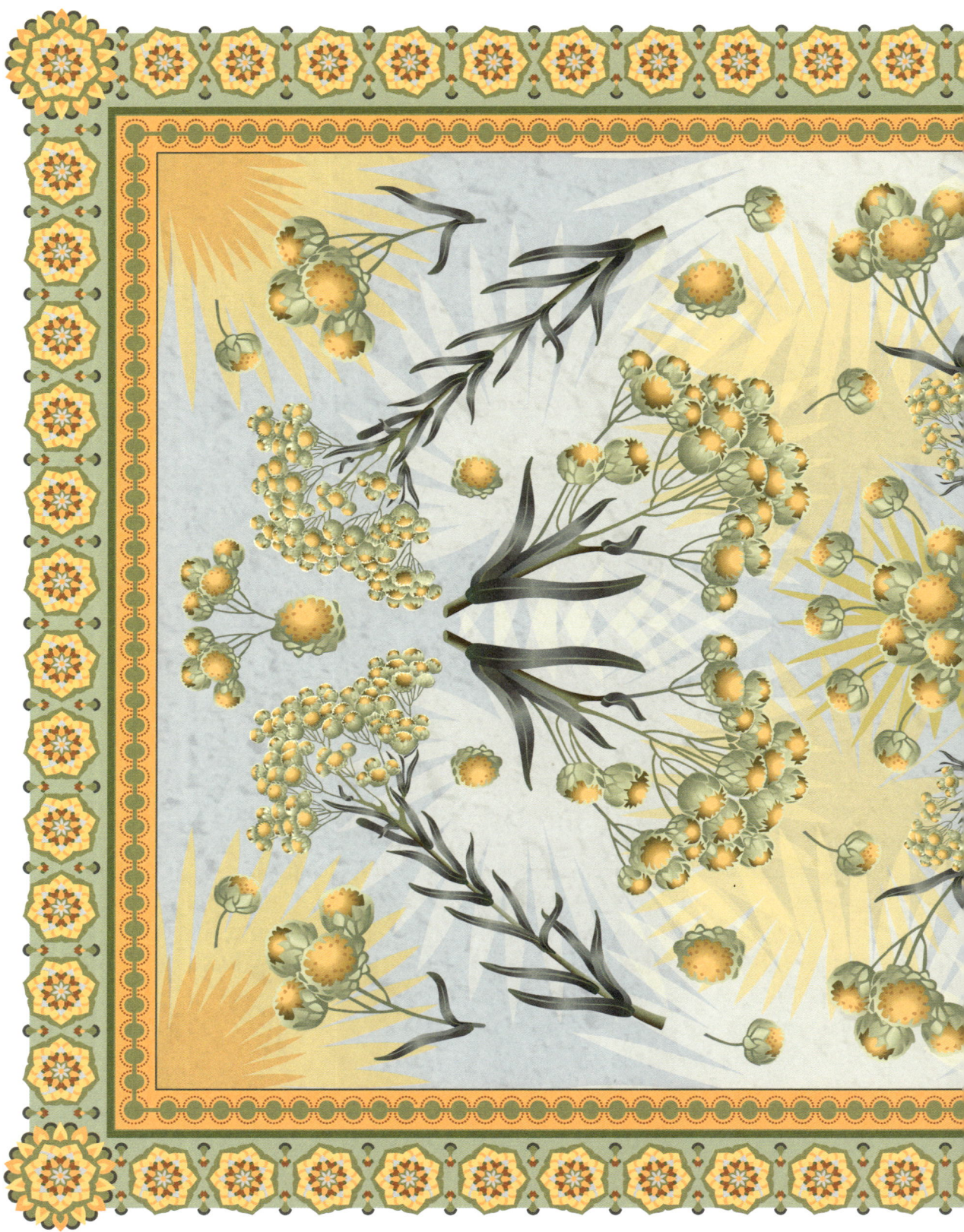

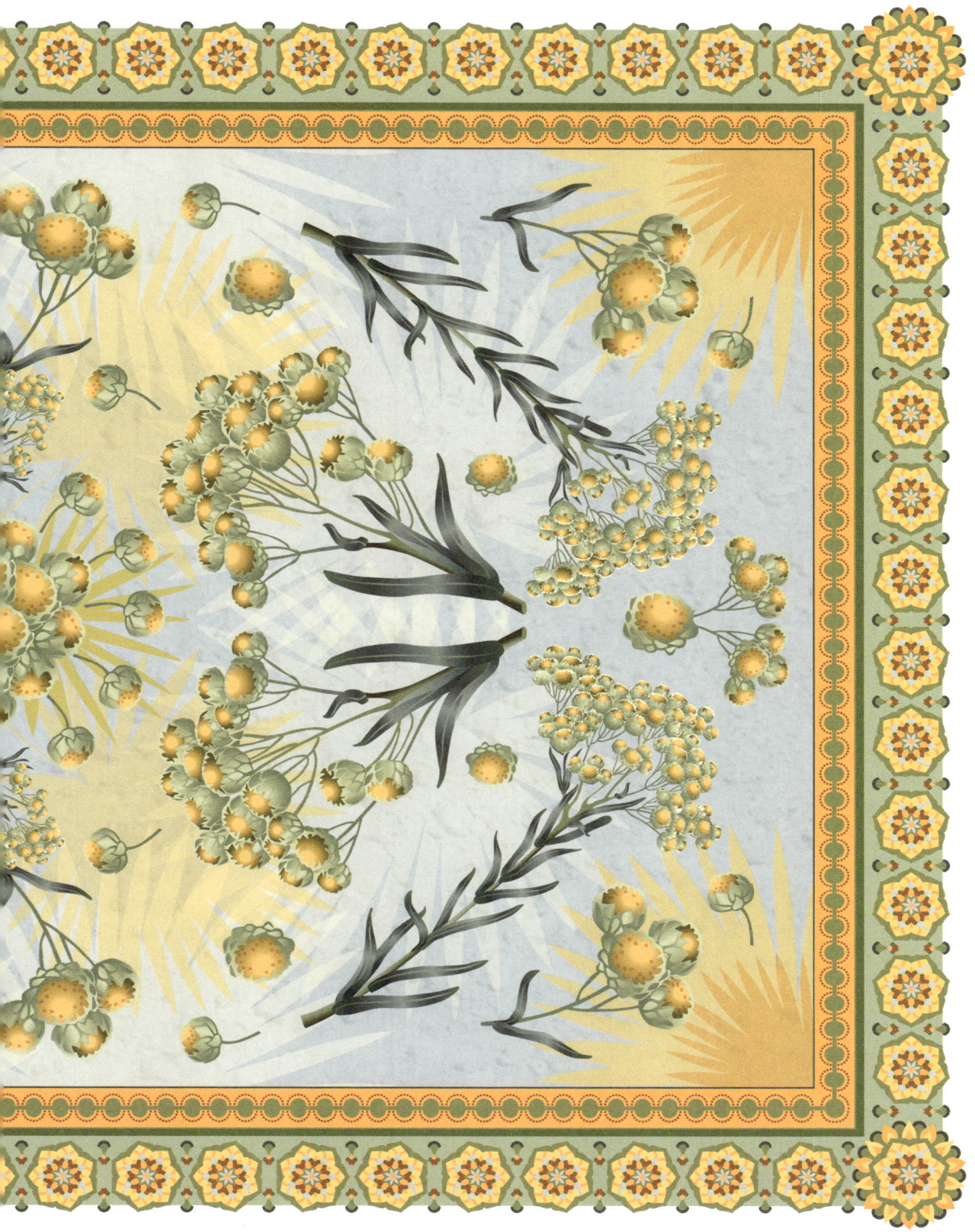

MAGNOLIA

Michelia alba

Readers of the first *Atlas of Perfumed Botany* will recall Madame Remy, a chemical engineer. She and her husband shared a passion for discovering and producing new raw materials for perfumery. In the 1980s, at China's invitation, she encountered osmanthus flowers in Guilin; working with local technicians, she developed a concrete that in turn was given final form in Grasse as an absolute. Over time, Madame Remy befriended her collaborators, who introduced her to another plant valued in China: the magnolia. Magnolias are primitive trees or shrubs in the *Magnoliaceae* family; comprising one hundred and ten species, magnolia appears to have changed little since its Cretaceous origins. Native to Asia and the Far East, magnolias are now found in the southern United States, Central America, and Europe.

In Guangxi province (Wuzhou), 500 kilometers from Guilin—a day's drive on a difficult road—there's a magnolia forest of 70,000 trees. Neatly aligned and growing up to fifteen meters high, they produce an overpowering scent when in bloom. Traditional bamboo ladders are used for harvesting. It's no mean feat to pick the flowers from that many trees. The operation requires acrobatic skill, and one needs to be young and flexible to carry it out. A ton of flowers, gathered one by one, yields twenty kilos of essential oil. The soil in the forest is good, so there are two harvests a year. The first, which is more abundant, takes place between May and June; the second occurs between August and November. Some of the flowers are dried and used to flavor teas, while others are used for producing essential oil to flavor tobacco. All that Madame Remy had to do was convince European perfumers to use the extract; its floral, rosy scent has notes of grapefruit and lavender, which is unique and surprising. Over time, essential oil from the flowers was combined with a less floral, greener extract made from the leaf, which has a slightly higher yield. Madame Remy offered these products to her clients in 1993. Master perfumer Maurice Roucel went to work and created Tocade for Rochas (1994). It was followed by the magnificent 24 Faubourg by Hermès (1995), Envy for Gucci (1997), and L'Instant by Guerlain (2003). It's rare for a new scent to be used so quickly; more often than not, it takes several years for an unfamiliar fragrance to show up in a perfume. More recently, perfumer Carlos Benaïm has created L'Eau de Magnolia (2014) for Frédéric Malle.

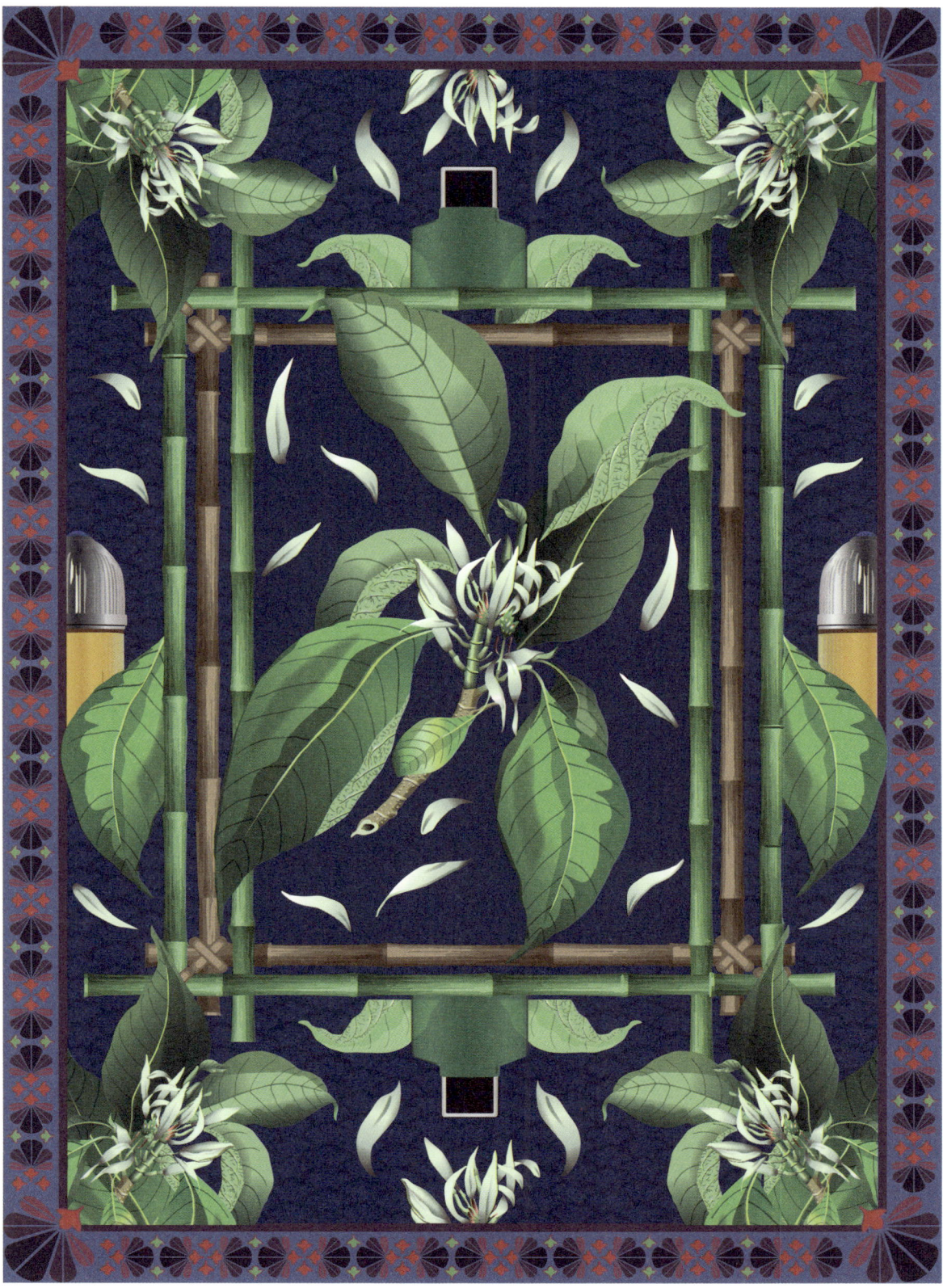

MARIGOLD

Tagetes patula, erecta et minuta

Ever since the Portuguese brought it from Mexico, the marigold has become the iconic flower of India. Marigold (*Tagetes patula*) and Aztec marigold (*Tagetes erecta*), with which it is often confused, belong to everyday life and to all celebratory occasions. There isn't a temple, hotel, restaurant, store, car, cab, or even rickshaw that doesn't have

In Mexico, marigolds serve as offerings on the Day of the Dead.

a garland somewhere. The Ambassador, India's signature motorvehicle, is decked out with wreaths of these flowers at weddings. In France, we throw rice to symbolize the wish for fertility and prosperity. In India, where marriage is more the union of two families than that of a couple, the custom is to throw handfuls of marigold petals as a sign of joy and wealth.

To appreciate the importance of this charming and brightly colored flower, we need look no further than the first few minutes of Mira Naire's *Monsoon Wedding* (2001), which offers a humorous portrayal of marriage preparations for the daughter of a wealthy Indian family in New Delhi. Or we can watch John Madden's *The Best Exotic Marigold Hotel* (2011), a British comedy in which a group of senior citizens leaves London to live in a restored palace for economic reasons; what awaits them there turns out to be quite different from what they expected.

In Mexico, marigolds, or *cempoalxúchitl*, symbolize just the opposite. Here, they serve as offerings on the Day of the Dead. One ritual involves creating a carpet of petals in orange, yellow, and red—the colors of a setting sun; it forms a path for departed souls to follow to the altar their loved ones have erected. The festival dates back to the time of Aztec civilization, when the departed were celebrated twice a year: first children, then adults twenty days later. When the Spanish arrived, the two events were moved to coincide more or less with All Saints' Day. In France, All Saints' Day is associated with chrysanthemums—the flowers of happiness and national celebration in Japan.

Tagetes are herbaceous members of the *Asteraceae* family native to tropical regions of the Americas from Bolivia to Mexico. A vast array of cultivars exists. Plants are bushy, branched, and upright, with highly articulated, symmetrical foliage. Marigold flowers measure two to five centimeters in diameter and occur in ones or twos; the flowers of Aztec marigolds take the form of pompoms five to twelve centimeters in diameter. Flowers are edible, and their yellow, orange, and red petals brighten up vegetable and fruit salads. These coloring properties have earned the plant the nickname of "poor man's saffron." The leaves emit a rather unpleasant odor, which is said to keep snails and other pests away. For this reason, tagetes are often grown in gardens. I myself have tried them out among tomatoes. The results weren't too great in terms of pest control, but the aesthetic effect was astounding and earned me a round of applause from my family. Apparently, I'd used the wrong cultivar. I should have gone for the *minuta* variety, which has nothing ornamental about it; it's slender and tall (two meters) and has late-blooming, small flowers. This is the kind that is distilled in India, Madagascar, and Egypt to obtain essential oil. It is used in limited quantities in perfumes, which makes it a rare but captivating essence. A few handfuls of leaves added to neroli oil (bitter orange blossom) boost the latter's scent. Tagetes is one of the "trace" products: components that are not analytically detectable but mark a fragrance. It is found in Karl Lagerfeld's Chloé

(1975) by perfumer Betty Busse. The vegetal, fruity character of this essence provides the perfect accompaniment to the floral notes of tuberose, as in Fracas, a perfume by Robert Piguet (1948). The same effect can be found in Ralph Lauren's forgotten Lauren (1978) and Eau de Givenchy (1980), a delicate composition in the Japanese style by Daniel Hoffman Perfumes.

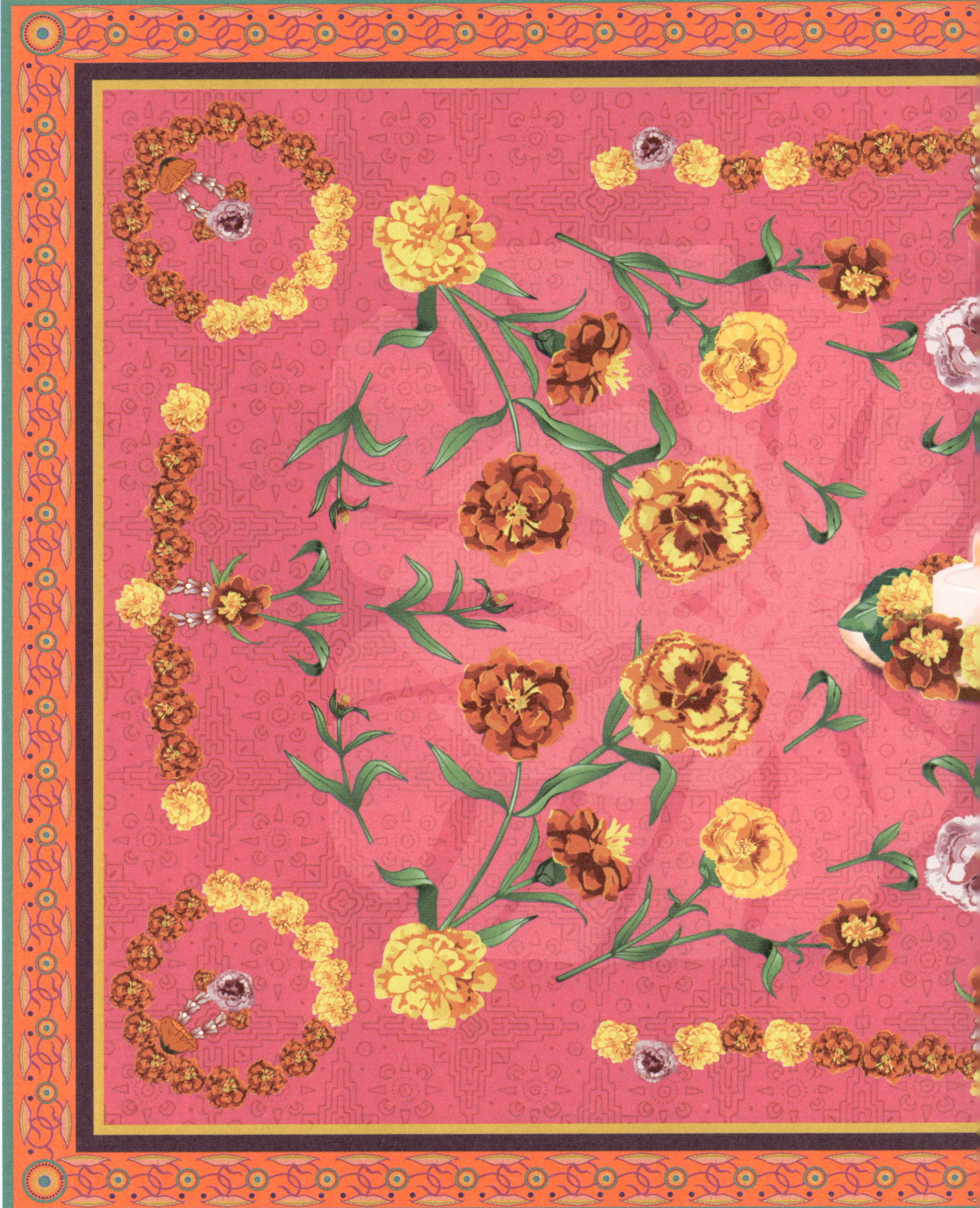

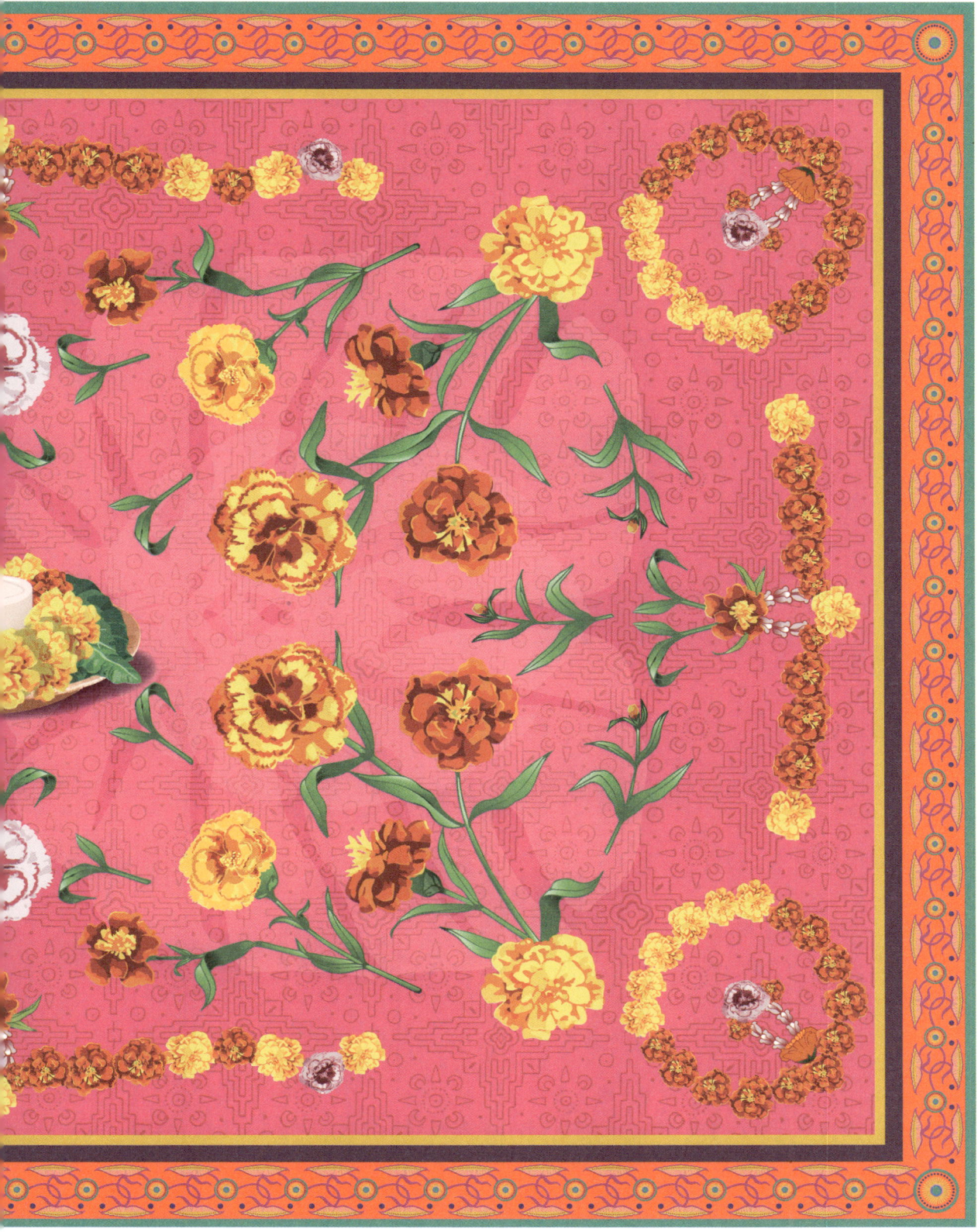

A plant product that results from the growth
of the flower and contains the seeds.
Littré Dictionary

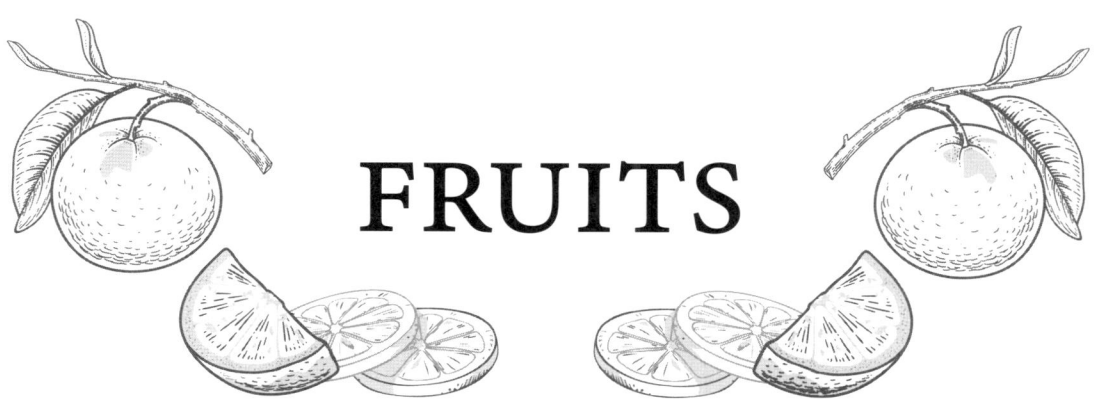

FRUITS

A journalist asked me whether perfumers use essences of caviar lemon, calamondin, kumquat, pomelo, cedrate, and other fashionable citrus varieties. I replied that, unlike Michelin-starred chefs with their vegetable gardens and orchards, which allow them to have all sorts of herbs, vegetables, and fruits at their disposal, perfumers' professional needs are counted in tons; consequently, natural extracts come mainly from what is most common and widely consumed. The scents of rare fruits, whose names abound in perfumery but from which no essence is known to be extracted, for the most part result from combinations of natural products and a few chemical molecules. To give you the picture of industrial production: each year, 7,000 tons of lemon essence are consumed, as are 900 tons of lime essence, 300 tons of mandarin essence, 120 tons of bergamot essence, and 10 tons of tangor essence. For cedrate and yuzu, the quantity is on the order of a few kilograms. To illustrate the size of the orchards involved, I propose a fifth-grade-level exercise in arithmetic: "If a lemon tree produces 400 kilos of fruit a year, if it takes an average of 200 kilos of lemons to produce one kilo of essence, and if a space of five meters must be left between each tree to allow it to grow, how big does an orchard need to be to yield 7,000 tons of lemon essence?"[8]

8. Solution: one tree produces two kilos of essence, and 500 trees one ton; that makes 2.5 kilometers of trees in a row, or 12.5 hectares of orchard, per ton of essence; multiply that by the 7,000 tons of essence produced yearly and one gets a growing area the size of the Bordeaux vineyards.

CEDRATE

Citrus medica

Infrequently seen in stores, this oval fruit has a disturbing appearance, like a rugby ball with dented leather. We most often encounter it as candied fruit used for decorative purposes, or in the form of Eau de Fleurs de Cédrat, an eau de Cologne created by Jacques Guerlain in 1920.

Although little known, it is in fact the oldest citrus fruit and the origin of the common lemon (*Citrus limon*); the latter comes from a cross between bitter orange, cedrate, and probably lime (*Citrus aurantiifolia*). Is this why the Jews call it the fruit of the "beautiful tree" mentioned in the Torah? The cedrate known as etrog is a variety used for Sukkot, or Feast of Booths, which celebrates the divine assistance the Children of Israel received as they fled Egypt to the Promised Land. During this autumn festival, the community is invited to live in a hut, or *sukkah*, and to celebrate its seven days with a morning prayer, holding a *lulav*—a bouquet of palm, willow, and myrtle branches—in the right hand, and an etrog in the left. A winter fruit, this cedrate honors the last harvest of the year. That's how important—and how ancient—it is.

The cedrate originated in the northeastern regions of India, Yunnan (China), and Myanmar (Burma). A thorny tree between two and three meters tall, it has the particularity of flowering year round. It is said to be the first citrus fruit to have been imported to the West. Theophrastus (372–288 BCE) described it in his study of plants. Buddha's hand (*Citrus medica var. sarcodactylis*) is an Asian variety; it is offered on the occasion of the New Year as a token of good luck. The first uses of cedrate for alimentation were described in the second century, and depictions of the fruit have been found on frescoes and mosaics in Pompeii. At the end of the nineteenth century, the alimea variety was intensively cultivated in Corsica for eating purposes; when exported, it served in the manufacture of candied fruit to flavor a uniquely Italian specialty: panettone. Today, the cedrate tree is grown mainly in North Africa around the Mediterranean basin, but also in China and Brazil.

For some time now, there has been a real craze for the taste and smell of this fruit. Cedrate can be found in fragrances such as Cologne Cédrat by Patricia de Nicolaï (2006), Cédrat by Roger & Gallet (2007), Cédrat Enivrant by Atelier Cologne (2013), Eau de Cédrat by L'Occitane en Provence (2015), and Incense & Cédrat by Jo Malone London (2015).

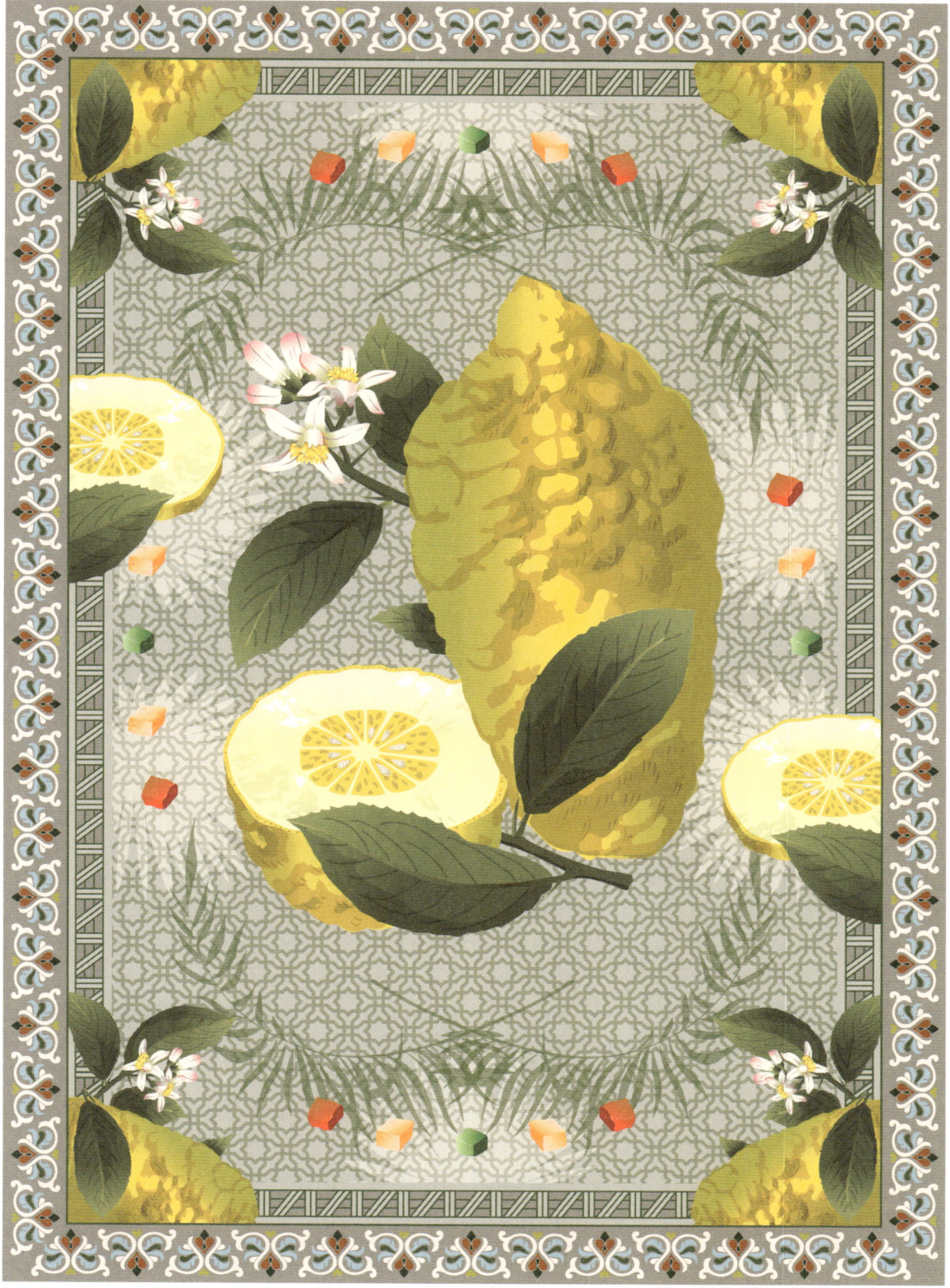

LIME

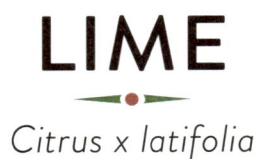

Citrus x latifolia

The lime tree is a shrub in the *Rutaceae* family, and there are two species: *Citrus latifolia* and *Citrus aurantifolia*, both of which are native to Southeast Asia. In the nineteenth century, *Citrus aurantifolia* was supplanted by Tahitian lime, which is a natural hybrid between a lemon tree and the Mexican lime tree. It is also known as Persian lime, although its Persian origin is disputed. Growers have opted for Persian limes because the fruit is twice as large and seedless; it also grows quickly, making it more productive and better able to withstand transport.

This small tree about two meters tall has a single, branched, and slightly thorny trunk. Its hard, lanceolate leaves are arranged in an alternating pattern, and they give off a strong odor when rubbed. The fruit is beautifully round, with juicy green pulp and a pleasant aroma. As is the case for all citrus fruits, the rind remains green in the tropics; it turns yellow in the Mediterranean. The lime was introduced to California at the end of the nineteenth century, then to Florida around 1930; subsequently, it came to South America and, more recently, India, Spain, and Portugal.

Today, Mexico and Brazil are the world's biggest growers of lime; for decades, Florida held first place. Together, these countries deliver 50,000 tons of fruit per year. Perfumers obtain the two essences used in their craft as a by-product of juice production. The first essence is distilled lime, or "limette," a misleading name because it refers to *Citrus latifolia*, which it is not. The second comes from cold pressing. The former calls cola drinks to mind, and the latter resembles lemon essence. The fruit's rind and juice are used for caipirinhas, mojitos, and daiquiris; other cocktails combine them with lemon, orange, cherry, strawberry, mango, rosemary, peach, raspberry, elderflower, watermelon, mint, coconut, guava, kiwi, mandarin, jasmine, blackcurrant, basil, cucumber ... Bartenders—or "mixologists," as they have come to be known—have no shortage of ideas to give traditional perfumers a run for their money.

Ideal for use in eaux de Cologne, lime produces an immediate effect of freshness. The acidity of its smell "grabs the nose." The first lime fragrance was launched in 1957 under the name Royall Lyme by the English company Royall Lyme Bermuda. In 1966, the Shulton Company in the United States introduced Old Spice Lime. Old Spice Original, dating back to 1938, remains one of the most widely used eaux de toilette in North America. Since then, many other fragrances have entered the market: Lime and Basil from Gandini 1896 (2010), Lime, Basil & Mandarin from Jo Malone London (1999), Limette 37 San Francisco from Le Labo (2013), Riviera Lime from Bella Bellissima (2014), and Limon Verde from Guerlain's Aqua Allegoria collection (2014).

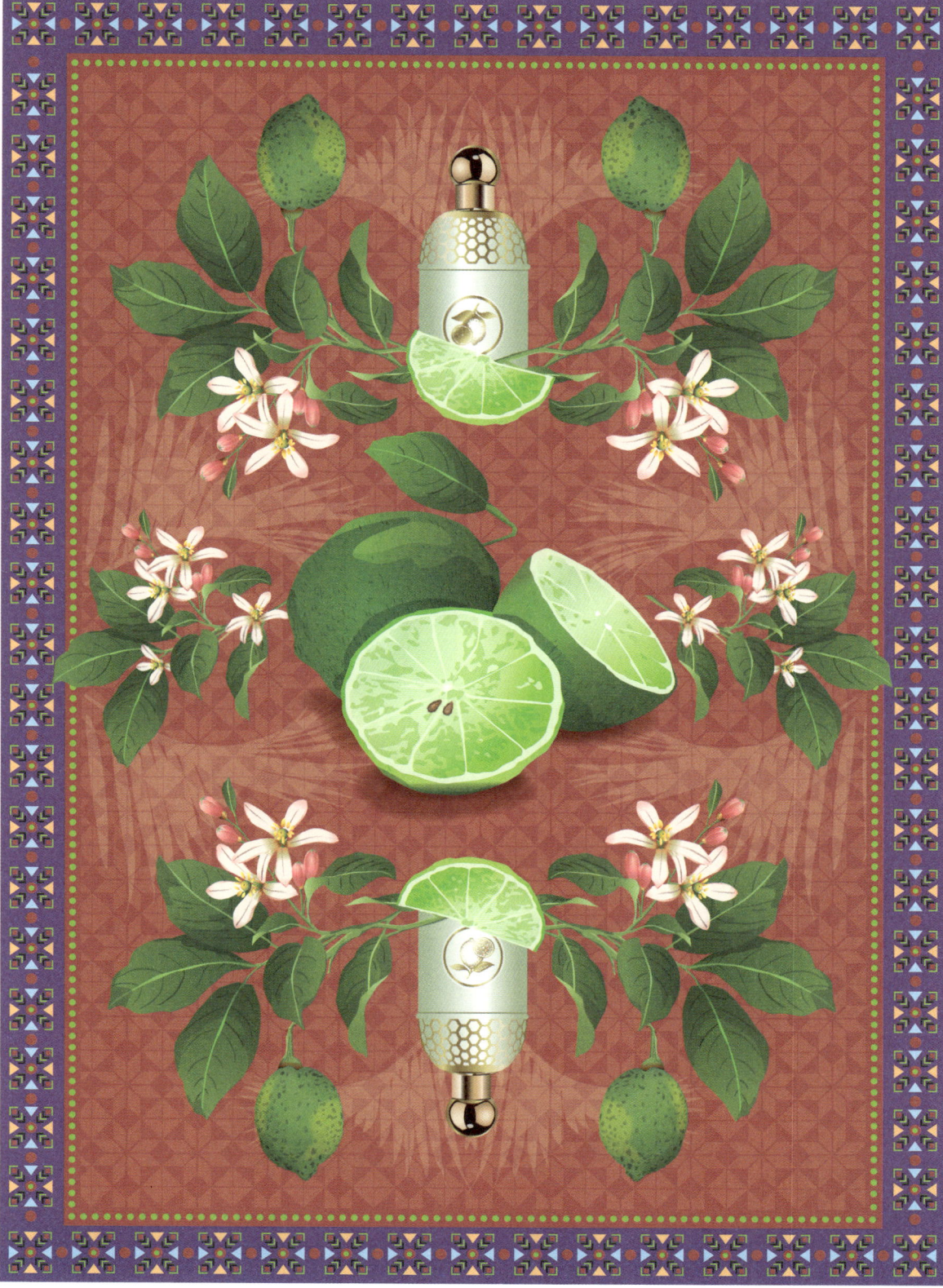

TANGOR

Citrus reticulata x sinensis

In France, Germany, England, Denmark, and Norway, *orange* is spelled "orange." It's the same word in all five countries; only the pronunciation changes. In Italy, orange becomes *arancione*. In Spain, it's *naranja*; the sound is similar. This brief linguistic survey shows that the orange, which is enjoyed the world over, has an appellation that's the same from one country to the next. This is not the case for tangor, whose common name is a contraction of *-or*, for orange, and *tan-*, for tangerine (so called because it came to the United States from Tangiers, in Morocco). Tangor is a word that cracks a smile—and lends itself to games of Scrabble. In Japan, it's also called tangor. The name in Israel is *topaz*, and in South Africa it's *tambor*; in Morocco it's called *ortaline*, in Uruguay *uruline*, in Spain *Villa Late*, and in Australia *australique*. Reading all these different ways of referring to the same thing is confusing. As the saying goes, "To pass undetected, change your name." Onomastic proliferation arises from the need to make the foreign one's own. To assign a name to something is to possess it; this is understandable in the case of this fruit, since no one knows its real parents—and everybody wants it.

Tangor was discovered in the market of Christiana, a small town in the parish of Manchester in central Jamaica. It is said to be the result of a chance cross between a sweet orange and a mandarin orange. The legend dates back to the end of the nineteenth century. A man by the name of Swaby noticed some unfamiliar fruit for sale and became curious; he bought two of them and planted seedlings from the pips. Some time later, in 1900, he presented the results at an agricultural exhibition, and they were selected as the most beautiful entry. Then he disappeared. Only decades later, in 1944, did the tangor reach the broader public, when the Citrus Growers Association of Jamaica started marketing the fruit for export.

The tree resembles a mandarin tree and bears up to 150 kilos of fruit per year at maturity. For a long time, Jamaica led in exports of this almost seedless and low-calorie but vitamin-rich and very juicy citrus; it proved especially popular in the United Kingdom and New Zealand. In 2017, the Citrus Growers Association of Jamaica ceased operations, but the story has continued. Tangor is grown on the island of La Réunion, where the fruit ripens between June and September; in California, where it matures from February to April; and, finally, in Cyprus, where the season extends from January to the end of June.

YUZU

Citrus junos Siebold ex Tananaka

Here's a kind of citrus that's little known in France but widely used in Japan; its name entered our dictionaries rather late. After making a splash in the culinary arts, it is now showing up in perfumes. Yuzu came to Europe in conjunction with the travel boom that took place from the 1970s on. Michelin-starred chefs' infatuation with Japan and its traditions led to so-called *nouvelle cuisine*. The movement, which started in France, received its title from the food critics Henri Gault and Christian Millau. When discoveries like this are made, experts in tastes often beat experts in scents to the punch. Originally from China, yuzu was introduced to Japan and Korea during the Tang dynasty, which lasted from 618 to 907. As is the case for mandarin, the essence used in perfumery depends on whether the fruit is green, yellow, or red; each has its own organoleptic particularities, ranging from acidic to sweet. Thought to be a hybrid between the Ichang lemon and Satsuma mandarin, yuzu is harvested either green or yellow, depending on the taste desired. The green zest is used in summery drinks or mixed with chillies and then ground into a purée to produce *yuzukoshō*, a spicy condiment used (sparingly) for miso soup or sashimi. The juice of the ripe yellow fruit is highly acidic—twice as much as lemon—and makes *ponzu*, which is a combination of vinegar from its juice and soy sauce. Yellow zest is used in pastries, sweet dishes, ice creams, and, more recently in France, chocolates.

The tree, which can reach five meters, has little fear of frost; it can handle lower temperatures than most citrus. The branches are rigid; the wood is dark and quite thorny. The leaves are lanceolate, and the flowers white. The fruit has thick, irregular skin and is small in diameter, from five to eight centimeters. Since it contains many seeds, yuzu has only a moderate juice content. Production of essence for perfumery started in Spain and does not exceed 100 kilos per year. It's the most expensive citrus essence on the market.

In France, enthusiasm for this novel taste has encouraged small-scale production of fresh yuzu for high-end restaurants. Perfumers have also fallen under its spell. Examples include Parfum d'Empire's Yuzu Fou (2008), Issey Miyake's L'Eau d'Issey Pour Homme Yuzu (2014), Acqua di Parma's Yuzu (2019), and Patricia de Nicolaï's Eau de Yuzu (2020).

Gum: a viscous, transparent substance
that comes from certain trees.
Littré Dictionary

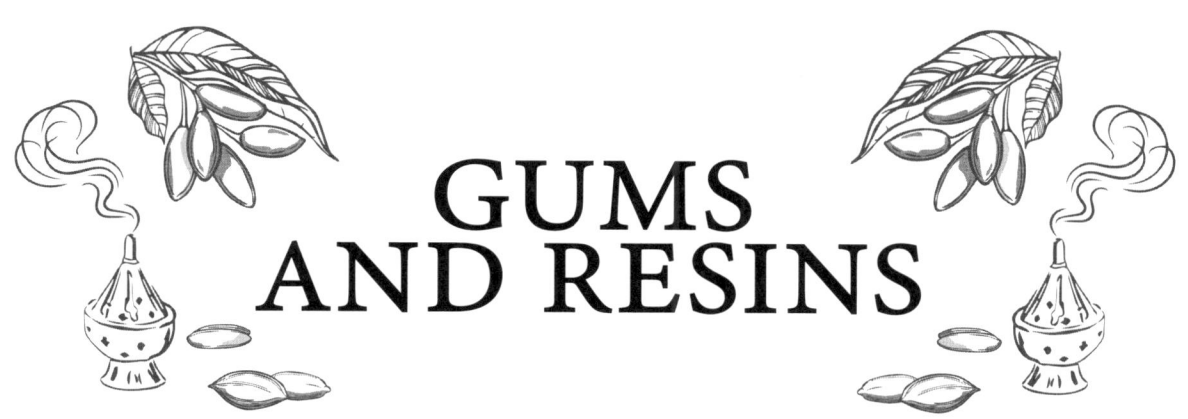

GUMS AND RESINS

Gums and resins are viscous substances, sometimes transparent but more often translucent or opaque, which flow from the trunks of certain trees and shrubs. Gums—for example, frankincense, myrrh, benzoin, and opopanax—solidify on contact with the air. Resins, in contrast, can be of varying thickness; they include labdanum, Tolu, balsam of Peru, and galbanum.

Five thousand years ago in Brittany, our ancestors placed fragrant bark and resins (mainly from the white birch tree) in earthenware bowls; they then burned these substances inside dolmens, that is, collective burial sites covered with stone and earth. The purpose of the fumigations was neither to make contact with the gods nor to disinfect the site (this notion did not exist), but simply to mask any stench that might be present when a body was laid to rest. It wasn't until a thousand years later that Egyptian priests began burning gums, resins, and rare and costly woods in order to honor the gods—a practice that all religions subsequently adopted. Today, fumigations are performed in many countries on the occasions of birth, marriage, and burial; in some African countries, they play a role in perfumes and erotic pursuits. In sixteenth-century Japan, fumigation enjoyed a promotion in status with the practice of *kōdō*, or "way of perfumes," one of the three traditional arts, along with *chadō*, the "way of tea," and *kadō* (*ikebana*), the "way of flowers."

ELEMI

Canarium luzonicum

Elemi gum originated in Ethiopia and belongs to the incense family (genus *Boswellia*). Long used medicinally, in the eighteenth century it became one of the secret ingredients of varnishes made by luthiers for violins and other stringed instruments. Here's a recipe published in 1773 from Jean-Félix Watin's *L'Art du peintre, doreur et vernisseur*: "Add to a pint of wine spirits [low-titration alcohol] four ounces of sandarac, two ounces of shellac in grains, two ounces of lentisque in tears, and one ounce of elemi gum; melt these gums over a flame and, once they have come to a boil a few times, pour in two ounces of turpentine. An instrument made for frequent handling requires a hard varnish; consequently, add only a small amount of shellac in grains, as more will make it mealy. Avoid too much turpentine; it heats up in one's hands; gum elemi makes it harden and compensates for turpentine at a lower dosage."

Elemi is extracted from a large evergreen tree that started out wild in tropical Asia and Africa. Since then, the plant has been cultivated in the Philippines and the Sunda Islands. When buds appear, incisions are made in the trunk, producing yellowish-white wax tears that harden on contact with the air. A mature tree produces four to five kilos a year. The gum is collected and steam-distilled to obtain essential oil, or extracted with volatile solvents to make a resinoid. In the nineteenth century, this costly product was replaced by a more affordable gum from Brazil; economies of scale are nothing new. The newcomer, which also belonged to the *Burseraceae* family, received the same name; in turn, it replaced by modestly priced Manila elemi, which comes from *Canarium luzonicum.*

The essential oil's lemon, lime, orange, and pink berry scents exercise a seductive effect, and its price is very affordable. Elemi essential oil goes well with scents that "pop," particularly those of the pink berries known as rose pepper (*Schinus terebinthifolius*) or false pepper (*Schinus molle*); the latter is often confused with the former, but it is inferior in quality. Elémì Pour Homme by Helan, a recent addition to the world of perfumes, provides a notable example.

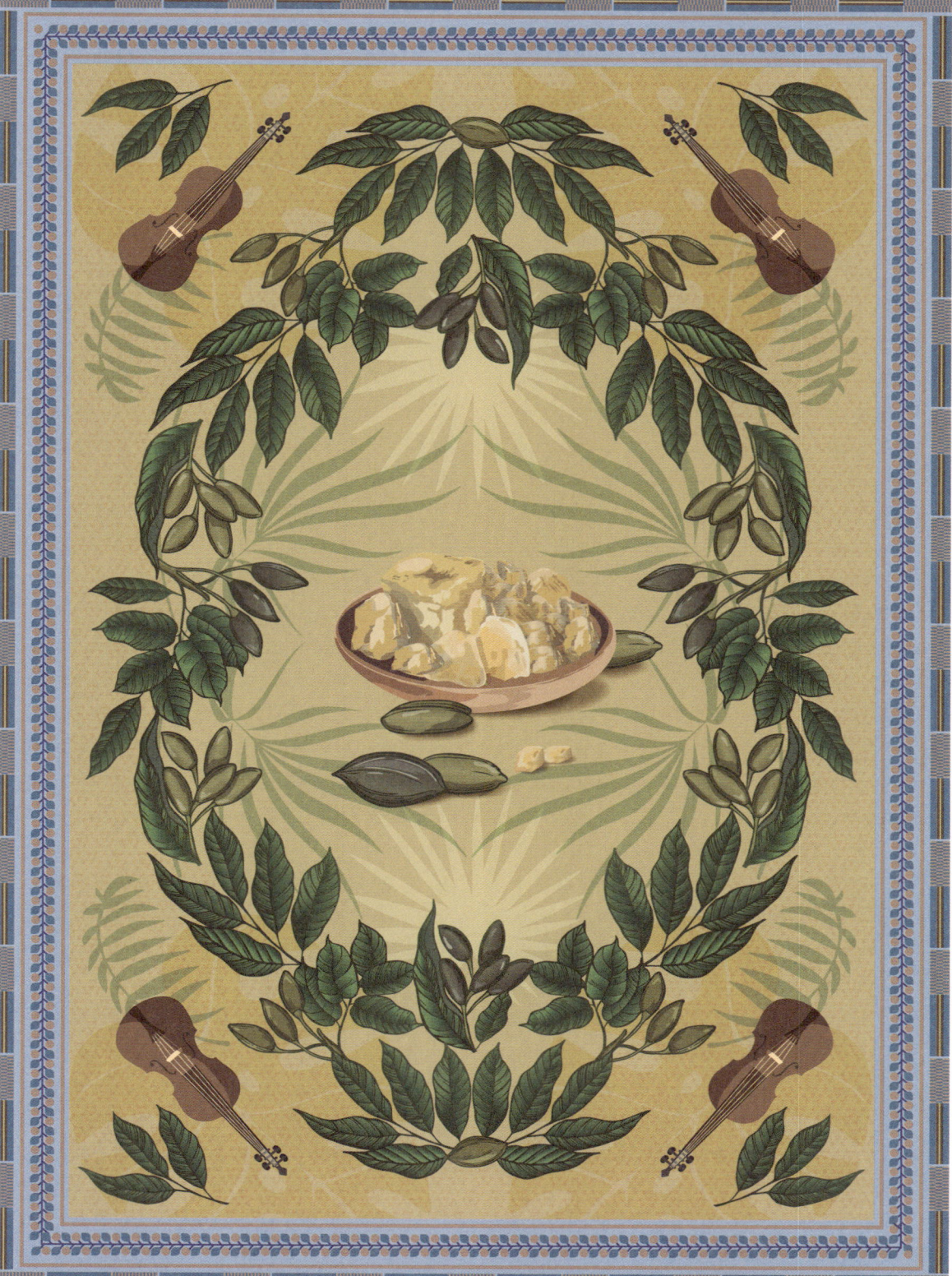

BALSAMS

BALSAM OF PERU

Myroxylon balsamum

Until the 1970s, it was a tradition for manufacturers of raw materials in Grasse to produce natural blends called "bases" at the request of Parisian perfumers. One that I remember in particular had a high percentage of balsam of Peru and was called Vanille 33; I never found out what the 33 stood for. It also contained vanilla absolute, benzoin balsam, Tolu balsam, tonka absolute, and synthetic vanillin. Secrecy, complexity, and savoir faire were the order of the day. This blend, which had the scent of vanilla and the consistency of caramel, was being manufactured for use in an important perfume. The secret was well kept. Locked in a safe, the formula came out only when an order came in. As I worked in the atelier, I had avidly read the document, which was written in pen with a flowing hand. I thought I'd unlocked the mystery of the perfume that would be making the news, but when I finally smelled it, I realized that the truth lay elsewhere: Vanilla 33 was just one of many components.

Peruvian balsam belongs to the *Fabaceae* family and is native to Central America, growing in Mexico, Guatemala, Costa Rica, Colombia, and Ecuador. Today, it is harvested only in the mountains of El Salvador and Nicaragua. Scientists call it *Myroxylon balsamum*. Its vernacular name is inaccurate, since the tree is not found in Peru. This moniker can be explained by the fact that balsam was once exported from the town of Callao, Peru's main fishing and trading port, not far from the capital, Lima.

The tree, which has a slender trunk and attractive gray bark, can reach twenty to thirty meters; its compound, imparipinnate leaves are alternating, with ten or so leaflets. Flowers are small, white, and clustered in the leaf axils. A liquid resin called "balsam" exudes from secretory canals in all parts of the tree. In his book *Cueilleur d'essences*, raw materials sourcer Dominique Roques describes the difficult and dangerous job of resin extractors, who work independently. These courageous individuals climb up trunks, meter after meter, and scarify the bark with a knife. To stimulate exudation, they place a burning torch over incisions and, with the aid of a makeshift fan, ventilate the wound without ever burning it; this leads to an outflow of balsam. Then they wrap strips of cloth around the cut to soak up the resin. After two or three weeks, the strips are collected and boiled in water to recover the balsam. As this economic resource depends on the good health of the trees, extractors display an almost religious devotion to them. The harvest is sold as is, or further processed by distillation or extraction with ethanol to make a resinoid. Balsam of Peru is mainly used in dermatology for protective lip balms. It has limited value for perfumery, as it can provoke allergies; regulations are sometimes contradictory. Its main components are, in order of importance, benzyl benzoate, benzyl cinnamate, cinnamic acid, and traces of benzoic acid and vanillin; the last two components, present in negligible amounts, are responsible for the almond and vanilla odors. Quantity doesn't equal importance.

TOLU BALSAM

Myroxylon balsamum var. balsamum

Tolu balsam, or Santos mahogany, is native to Mexico and the Amazon rainforests of Colombia, Venezuela, Peru, and Brazil. The history of its name is similar to that of balsam of Peru, with the difference that it came from Puerto Compas Tolù, a port on the shores of the Caribbean in Colombia. This tree is slow-growing and can exceed twenty meters in height, with dense evergreen foliage and clusters of white flowers. The wood is commonly called balsamo; it has a "clean" appearance and is knot-free and resistant to fungal decay. In the eighteenth century, it gained favor with craftspeople who gave it the name "eye of vermeil" for its color, which varies from orange-yellow to crimson-

pink. Like rosewood, which was imported around the same time, it found use primarily in veneering. In the nineteenth and early twentieth centuries, it was used to make copies of eighteenth-century furniture, as well as for interior woodwork. Today, it is widely grown as a shade tree on coffee plantations.

Tolu balsam comes from the tree's trunk, which exudes a semi-liquid resin that is darker than reddish-black balsam of Peru. The composition is similar, with one notable difference: the trace presence of eugenol, a spicy note that is also a major component of essential oil of cloves. Like balsam of Peru, and for the same reasons, Tolu balsam is used in cosmetics and dermatological products. Perfumers use it for the composition of vanilla and amber. These odors, which lack character on their own, often act as a binder or to conceal a few flaws, just as fat and sugar do in cooking. Tolu balsam, like balsam of Peru, was long sold as an infusion or tincture, that is, it was diluted hot or cold and macerated for several months in ethanol. These scented alcohols served as carriers for perfumes until the nineteenth century. At the beginning of the twentieth century, a new tax on the transport of spirits, known as the "excise duty," increased the price and put an end to their manufacture and trade.

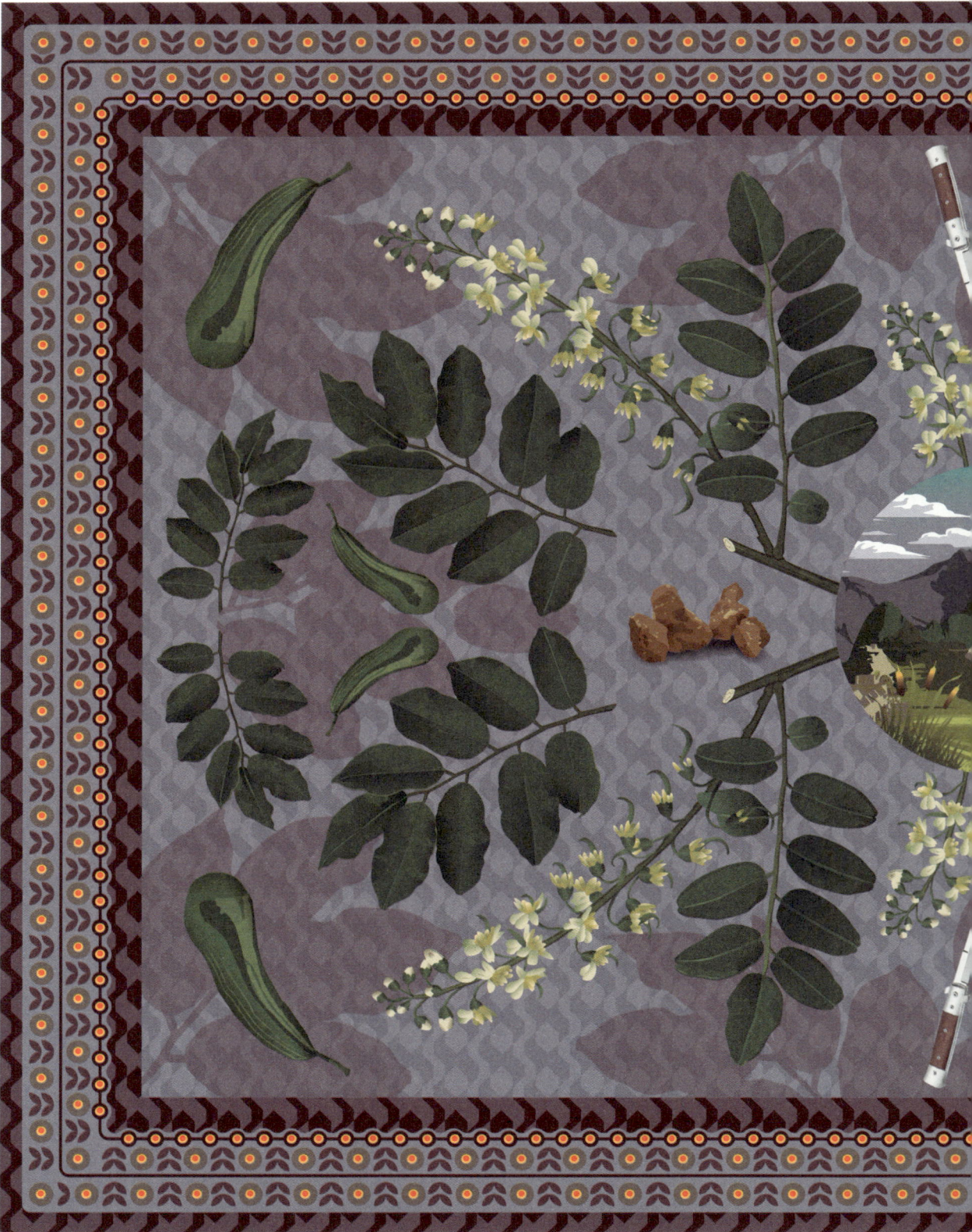

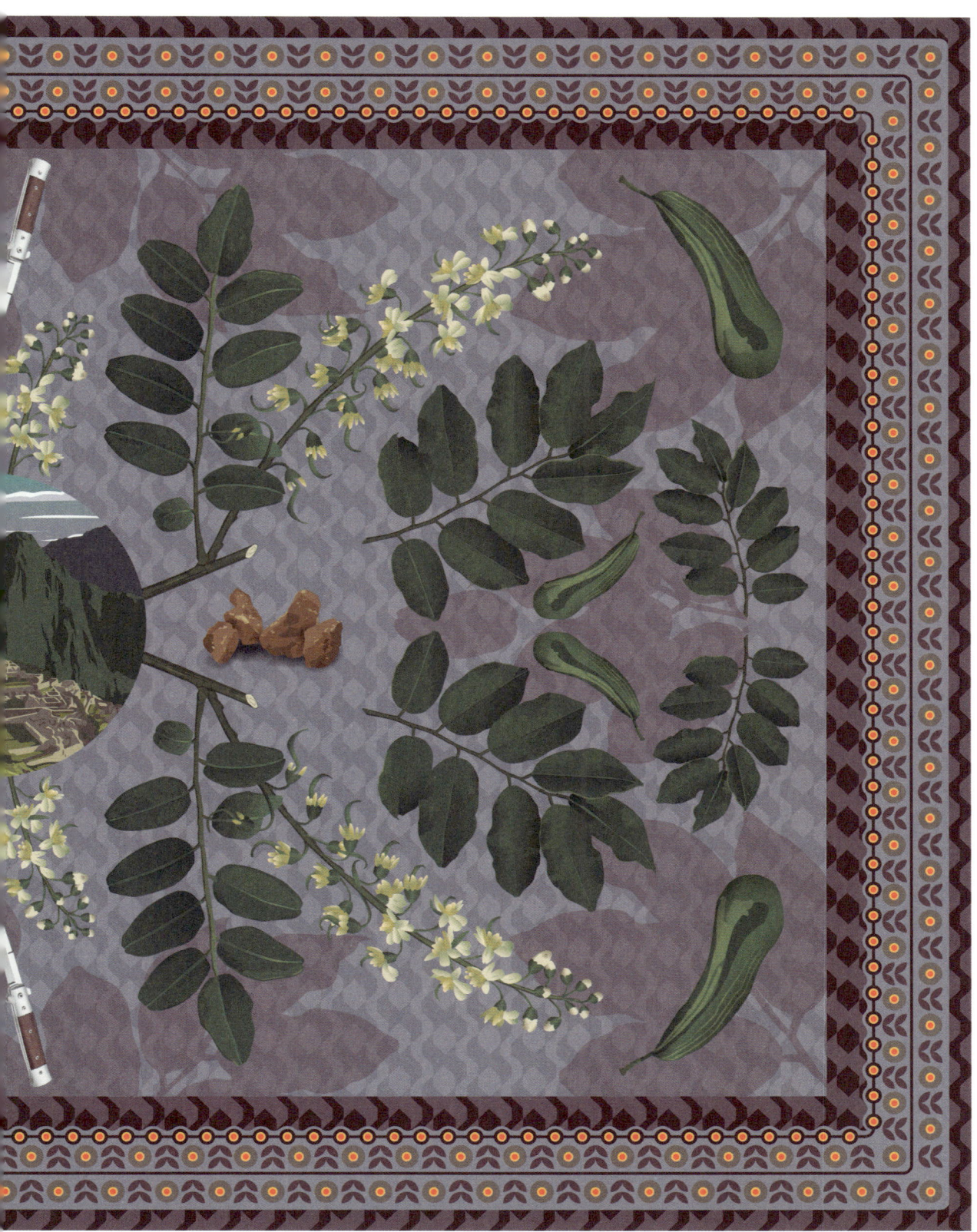

OPOPANAX

Opopanax was one of the products—and not the only one—that I was going to use in my early days as a perfumer without really knowing what it could contribute. I liked the smell, but I couldn't feel the effect. It was like salt that doesn't add salt. All the same, I still have respect for it.

Opopanax, which means "all-curing juice" in Greek, has been said to come from the root of a herbaceous member of the *Apiaceae* family related to galbanum, which was used in medicine and perfumes in ancient Greece and the Roman Empire. Other sources assign this resin to *Commiphora*, a botanical genus belonging to the *Burseraceae* family that includes 185 species of thorny trees and shrubs growing in the Middle East and Africa—mainly Somalia, Ethiopia, and Eritrea. The second origin strikes me as more likely. Its gum resembles that of myrrh (*Commiphora myrrha*), the only difference being its golden-yellow color; myrrh is reddish-brown. As for the scent, myrrh reminds me of the bitter smell of rain on the cobblestones of Paris, while opopanax has a candied lemon smell, which perfumers value for making the fragrance of citrus fruits more lasting.

Confusion persists, however, because opoponax is also said to come from *Commiphora erythraea*, *Commiphora guidottii*, or *Commiphora gileadensis* (alternatively, *Commiphora opobalsamum*), so-called Mecca balsam, or balsam of Gilead. This balsam is mentioned in the Bible and owes its name to the region along the Jordan River; some botanists attribute its origin to *Pistacia palaestina*, or Palestine terebinth. Today, knowing where opoponax comes from remains difficult. Hosts of small harvesters provide small quantities of gum of nonstandardized quality. Sourcing is difficult, and tracking impossible. Purchasers sort the product by smell and appearance, declining 10–15 percent on average.

In his *Natural History*, Pliny the Elder mentions opoponax as one of the ingredients of the "Royal Perfume" of the Parthians; this scent has been reconstituted in Versailles at the Osmothèque, the world's only perfume conservatory. Opoponax features in Santa Maria Novella's Opoponax, Christian Dior's Poison (1985), Yves Saint-Laurent's Opium (1977), and Tom Ford's Noir (2012).

My perplexity remains.

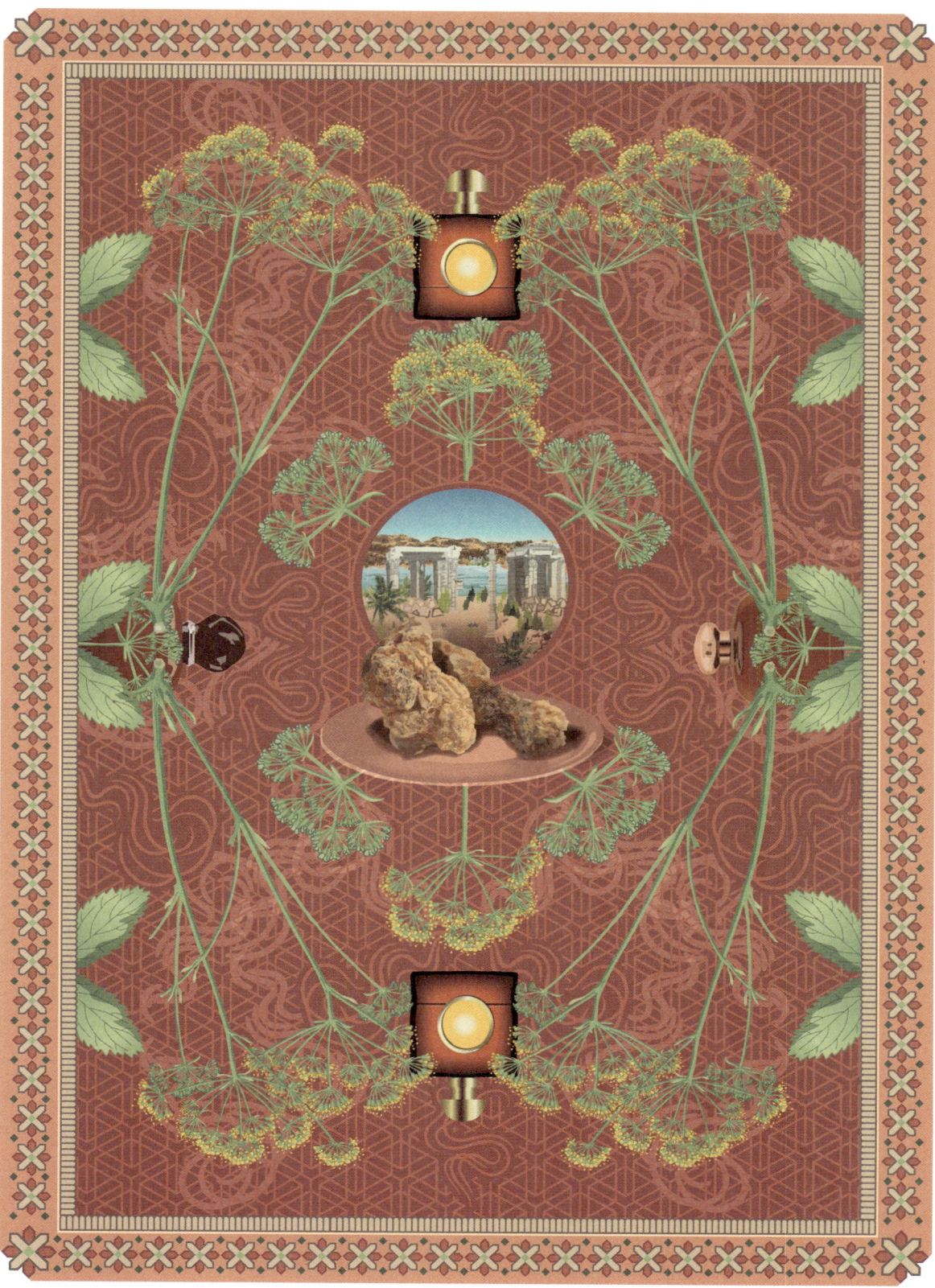

Fruit is the seed of cereals.
Littré Dictionary

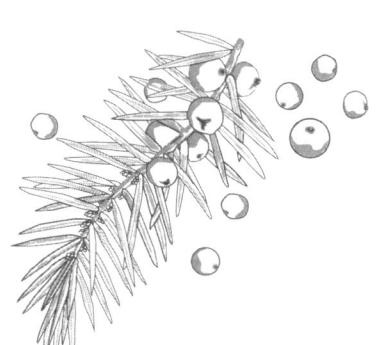
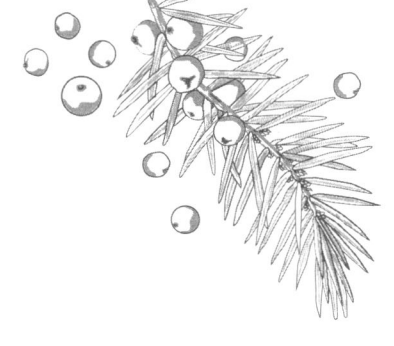

SEEDS

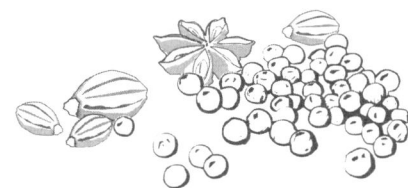

Jean-Jacques Rousseau was right: "If plants are given names other than the popular ones, those who gather them in the countryside, the herbalists and druggists to whom they take them, and the doctors who prescribe them, will no longer understand each other, and this confusion of languages will have unfortunate consequences." The same could be said about the perfume industry, which sometimes lacks precision in what it calls essential oils. The seeds of the umbellifers that have recently been named *Apiaceae* prove to be dried fruits, or mericarps. I have enormous respect and admiration for them. Encased in a woody carapace that protects against the heat and cold, they are life in the making. Perfumers and distillers of gin, vodka, sake, and whiskey have recognized as much and turned it to their advantage. They buy seeds when harvests are good, that is, at low prices, hold them at the ready for years, and distill them when the market's right. They're gold that grows.

STAR ANISE OR BADIAN

Illicium verum

In my opinion, star anise is one of the prettiest seeds, with its eight woody capsules, each of which encloses a grain; the other is ambrette, which resembles a small snail. The Chinese, who don't lack imagination or a sense of poetry, call it "eight-horned fennel." As a perfumer, I think of sixteenth-century pomanders, or *pommes d'ambre*: works of gold with silver chasing that open up into four or more receptacles for holding different fragrant pastes.

The badian tree, with its white trunk and pyramidal crown, is native to China. It resembles a magnolia with small lanceolate leaves. This point of similarity was enough for it to be called magnolia for a long time, even though it doesn't belong to the same botanical family. It is medium-sized in North America but can reach up to eighteen meters in subtropical zones. Its leaves are evergreen and glossy, and the flowers form clustered stars of pale yellow to pink. Fruits are picked green and turn reddish-brown when dried in the sun.

Returning from one of his voyages in the thirteenth or fourteenth century, Marco Polo introduced star anise to Europe. The shape of the new spice astonished Venetian merchants. Its taste was more familiar, resembling that of green anise (*Pimpinella anisum*) and fennel (*Foeniculum vulgare*), two *Apiaceae* that are common in Europe, especially along the Mediterranean perimeter. These two plants are prominently listed in Charlemagne's *Capitulare de villis*, a document addressed to city governors that recommends which herbs to grow in the gardens of abbeys, monasteries, and convents throughout royal dominions.

Coming from a distant land that kept secrets to itself, star anise was long considered precious and sold at premium prices. It achieved popularity at European courts thanks to its use in confectionery, cakes, gingerbread, and liqueurs. The real boom came in the twentieth century, when pastis was invented in Marseille. This occurred in response to the 1915 ban on absinthe; the so-called green fairy enjoyed great popularity, but the government condemned it for causing alcoholism. In 1918, Jules-Felix Pernod registered the Anis Pernod trademark; his beverage featured star anethole, which has a sweet taste free of bitterness. The drink's alcohol content was low; it was wartime, and a 16 percent limit had been imposed. Initially, the population shunned the product, because adding water didn't make it cloudy; a minor detail, but the ritual mattered. Eventually, under pressure from the public and local distillers, the law was relaxed. In 1932, Paul Ricard in Marseille introduced a clear herbal aniseed drink that was close to forty proof and, with the addition of water, clouded like the green fairy. History repeats itself, and with the outbreak of the Second World War, alcoholic beverages over sixteen proof were banned again to keep soldiers fit for combat. In 1951, Pernod introduced Pastis 51; this time, the beverage had an alcohol content of 45 percent. Success led to success, and other manufacturers made their own versions, incorporating different plants. In 1975, Pernod and Ricard merged. Incidentally, the quays of any large port (and especially Mediterranean ports) provide many strange sights; they have funny tastes and smells, too. For some time now, bars in Marseille have offered *mazout*, or "motor oil," a pastis drowned in Coca-Cola. As the French saying goes, "All tastes are found in nature."

The molecule common to green anise, fennel, and star anise is anethole, which is what gives these plants their taste and scent. It is obtained by distilling star anise in particular. Harvests occur twice a year, in April and October. Ninety percent of the star anise used worldwide (40,000 tons per year) comes from China's Jiangxi province.

In Europe, anethol goes into anisette and pastis in France, ouzo in Greece, and sambuca in Italy. Bakers use star anise, green anise, and fennel seeds to flavor cakes, gingerbread, candies, licorice, and pharmaceuticals. Essential oils of star anise, green anise, and fennel, as well as natural or synthetic anethol, can be found in Paco Rabanne Pour Homme by Paco Rabane (1973), Opium Pour Homme by Yves Saint-Laurent (1995), and Lolita Lempicka by Lolita Lempicka (1997). More recently, Brin de Réglisse (2007), from the Hermès Hermessence collection, has offered an an ode to lavender at the heart of winter—when the plant takes up position in tight gray cohorts and marches across the undulating fields. Notes of aniseed sometimes take the place of tarragon or basil, which find only limited use in eaux de toilettes and perfumes.

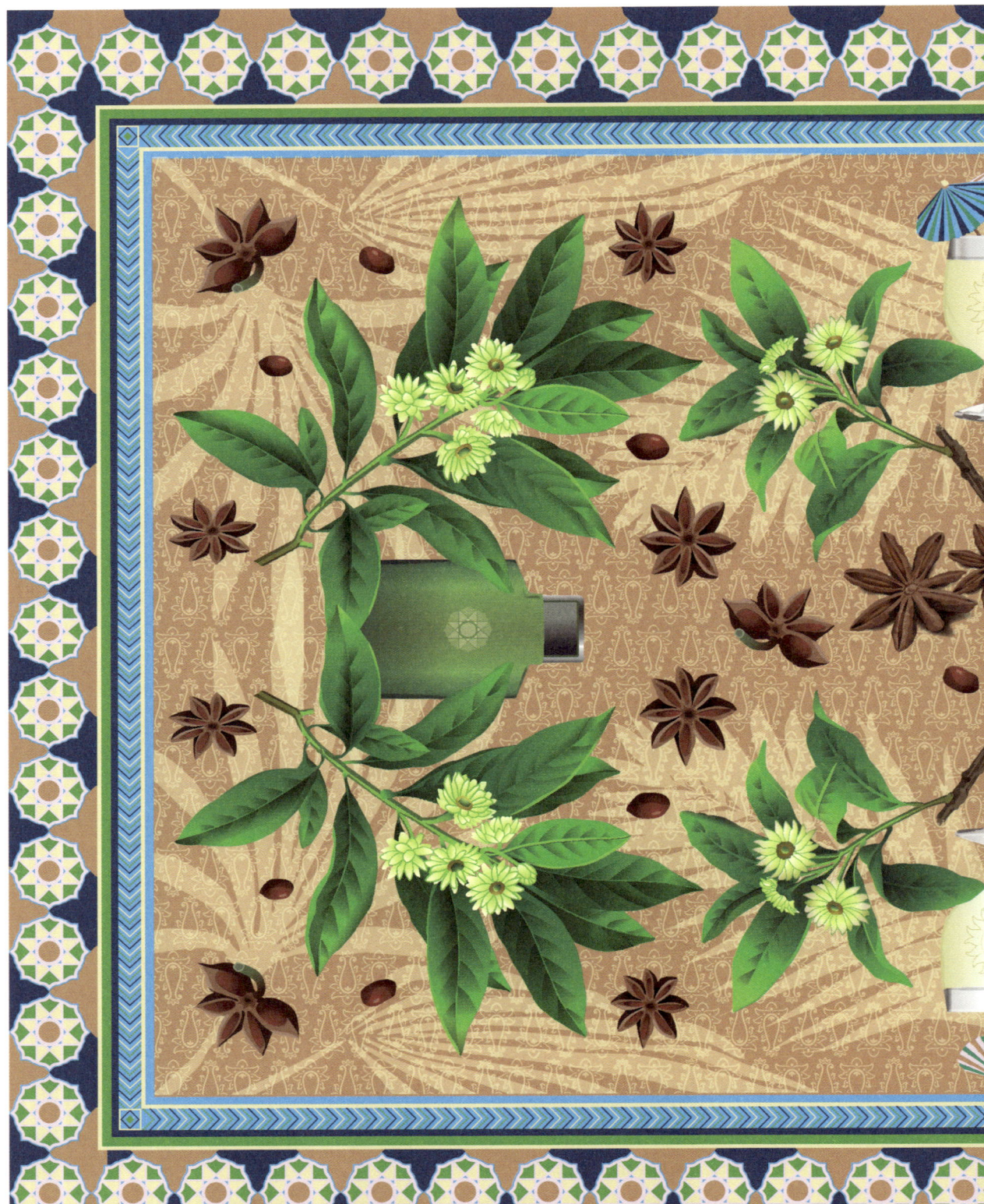

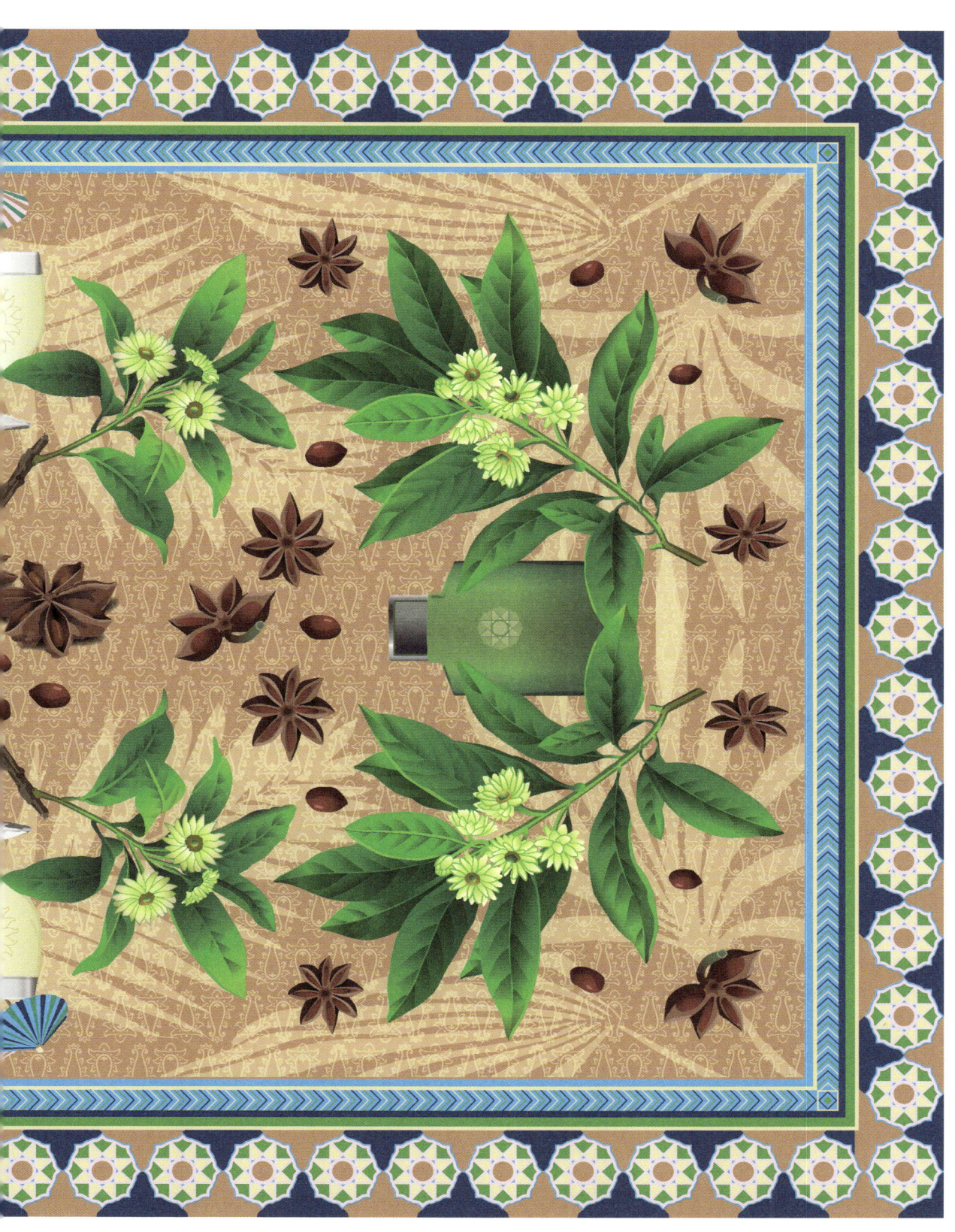

CORIANDER

Coriandrum sativum

This beautiful annual herb belongs to the *Apiaceae* family, formerly *Umbelliferae*. I prefer the latter designation, which comes from the Latin *umbella*, or "parasol"; it makes it easy to connect the plant with its kin. Jean Giono likened the stars to carrot flowers in *Joy of Man's Desiring*.[9] The carrot is another umbellifer, like fennel, parsley, or hemlock. Hemlock was given to those condemned to death in ancient Athens—most famously, the philosopher Socrates. It resembles parsley, but it is easy to tell the two plants apart on account of hemlock's red-stained stems, leaves that smell like cat urine, and highly toxic, green fruit.

Coriander leaves rubbed between one's fingers smell not like cat urine but like crushed bedbugs, which also isn't for everyone. An aromatic plant, coriander is easy to grow on a windowsill; just drop a few seeds into fresh potting soil and water it in the same way you would your parsley. Known as cilantro, the leaves are used fresh in culinary preparations; often they are confused with so-called Vietnamese coriander or *run-ram*, which has a very different taste. Cilantro essential oil's olfactory characteristics derive from one molecule in particular: trans-2,4-decadienal, which has a powerful lemon, orange, and bug scent; when used in trace amounts—just micrograms, which chromatographs are at pains to detect—it brings out citrus notes.

Essential oil from coriander leaves rarely factors into perfumery. It's a different story for essential oil from the seeds: with a linalool content between 60 and 70 percent, it unfolds the fresh, flowery smell found in rosewood and bergamot, making it a much-used component in fragrances. In my pantheon of scents, coriander seed numbers among the cold spices, along with cardamom, nutmeg, and pink berry. Each one of them adds a "lift": freshness, brightness, and an uncommon liveliness. But unlike the others, coriander doesn't have a scent that is characteristic enough to be identified in a perfume. It's there without showing itself, unless the perfumer makes a point of declaring its presence. Coriander is used in fragrances such as Coriandre by Jean Couturier (1973), Paloma Picasso by Paloma Picasso (1984), Knowing by Estée Lauder (1988), Féminité du Bois by Serge Lutens (2009), and Coco Noir by Chanel (2012).

9. *Que ma joie demeure* (Gallimard, 1935).

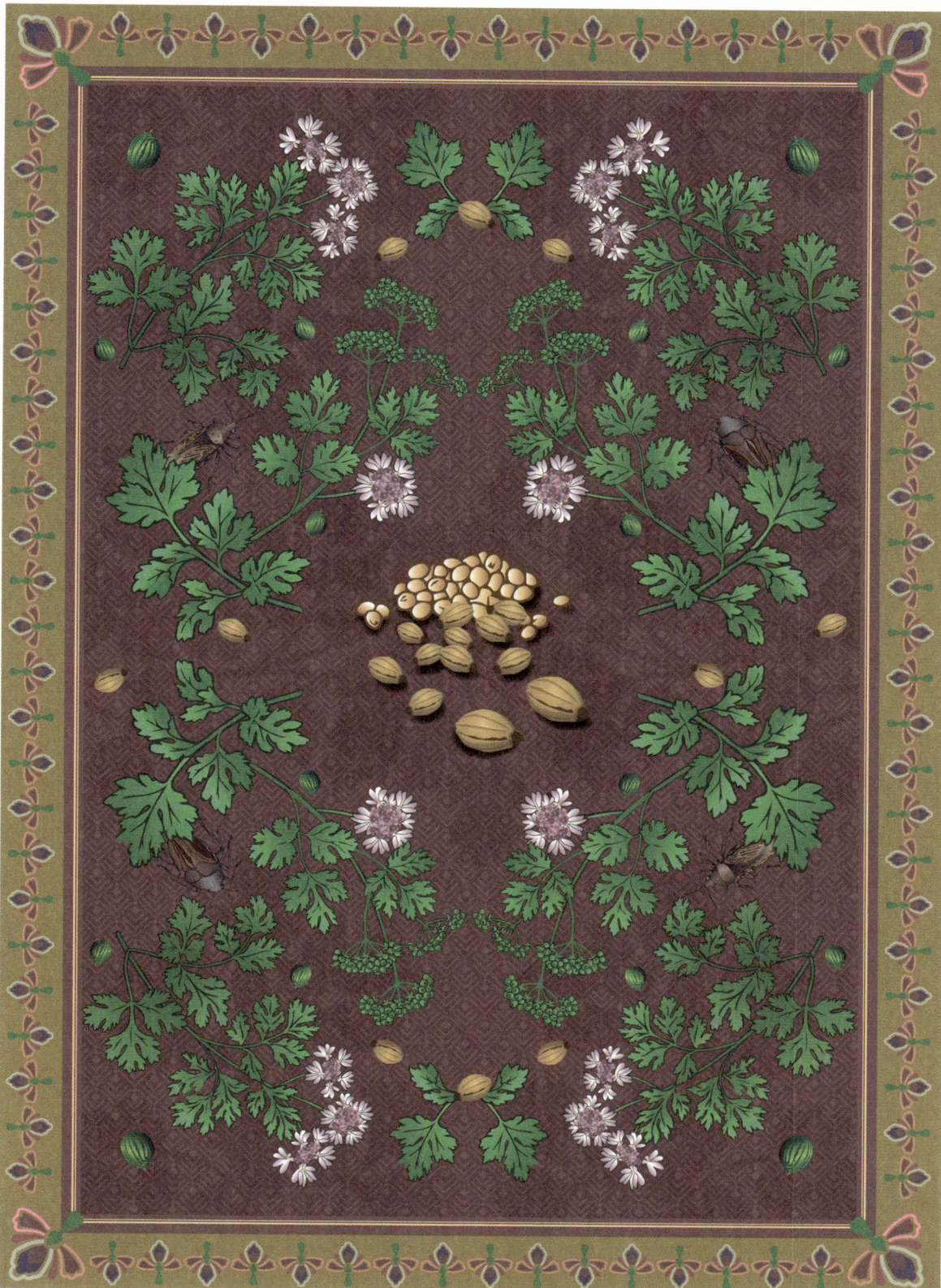

JUNIPER BERRIES

Juniperus communis

The trademark name "Voyage" had been registered. It was a perfect fit for the company, which would primarily be producing bags, luggage, and suitcases. We had agreed that the fragrance to accompany the line should be as much for women as for men, as travel has no gender. Émile Hermès had thought in a similar fashion in 1914, when he discovered the zipper—or "lightning system"—in the United States and obtained exclusive rights for its use in France.

> Harvesting takes place between July and September, when pickers beat the plant's branches with sticks.

An in-house craftsperson and artist received the commission of designing the bottle. The results were masterful. Inspired by a pocket-sized retractable magnifying glass found by chance on a trail, the bottle folded into a metal sheath for protection. As a matter of principle, I avoid seeing the containers for future perfumes, so as not to be prejudiced by sight—that sense which overrides the others. But this time, I had seen it. Now I had to add something that was off the beaten track—literally. I decided to create an accord of cold spices that would be inciting and lively; juniper berry essence would feature prominently at first, but amber would provide a comforting base.

Native to the Mediterranean basin, juniper (*Juniperus communis*) belongs to the *Cupressaceae* family of conifers. A thorny, perennial evergreen shrub, it grows wild at altitudes of up to 2,500 meters, mainly in the Balkans. Its height can vary from four to ten meters. The highest-quality berries come from France and Italy. Harvesting takes place between July and September, when pickers beat the plant's branches with sticks. The ripe, blue-black berries fall onto sheets spread below, while the green berries remain where they are, to be harvested the following year. They are then dried in the sun before being sorted. The best berries are used for cooking. Broken or irregular ones are used in perfumery; steam distillation begins in October and lasts for several months. It takes around 100 kilos of berries to obtain 1.5 kilos of essential oil. The oil is highly terpenic, so rectification is necessary to make it soluble in ethanol. Distillation of branches and twigs produces an essential oil of lower cost and poorer quality, with a scent similar to that of pine needles.

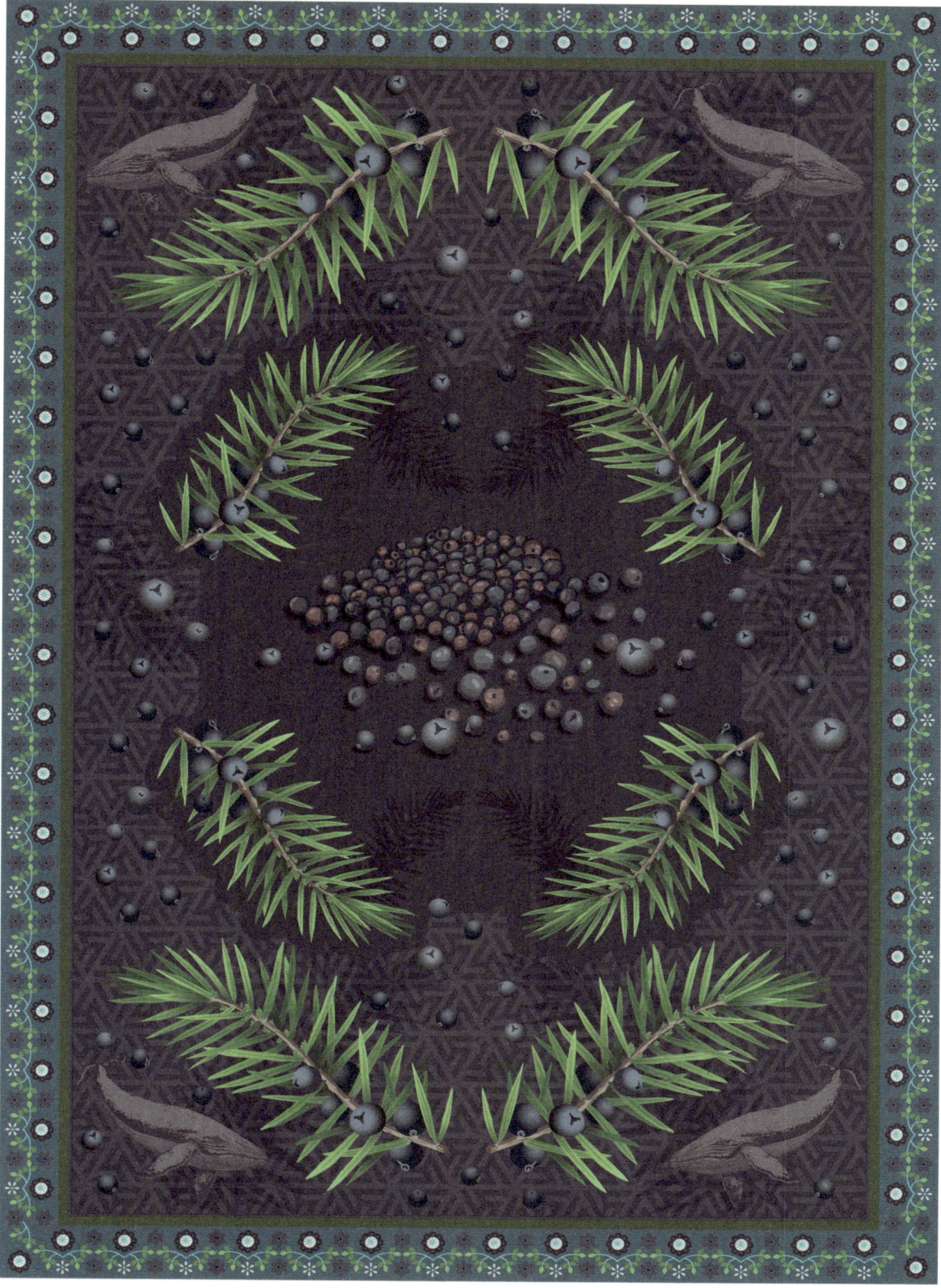

It's impossible to talk about juniper berries without mentioning gin. The ancestor of this easy-to-produce spirit, which is mainly flavored with juniper, is jenever—a favorite tipple among Dutch whalers who, until the nineteenth century, set off on voyages in the northern seas that could last for years at a time. Gin was born in the so-called Glo-

The inventiveness of mixologists knows no bounds.

rious Revolution, when the English ousted their Catholic king, James II, and Parliament called to the throne William III of Orange, Prince of the Netherlands and husband to Mary II, the deposed monarch's daughter. The Anglo-Dutch alliance stood opposed to France; wines and eaux-de-vie were taxed to the point of being unaffordable. In 1689, William III rescinded the state monopoly on spirits, allowing anyone so inclined to engage in the commercial production of inexpensive alcohol. Gin stores opened on every street corner—London alone counted over 7,000. Consumption rates soared in the young,

working-class population; the potation was cheaper than beer and an easy way to forget about one's difficult life. Gin became known as Dutch gin in England, Peket in Belgium, and Aquavit in Scandinavia.

Today, gin is fashionable. It stands far behind whiskey, which is available in 2,700 forms worldwide, but well ahead of vodka. Gin comes in some 400 iterations, including one called Comte de Grasse, which claims to be inspired by the craft of perfumers. It is first and foremost a cocktail spirit, containing between 37 and 57 percent pure alcohol. The best-known drink made with it is the gin and tonic, followed by the negroni, dry martini, and gin fizz; there are dozens more. The inventiveness of mixologists—cocktail craftspeople who focus on the quality of ingredients, balance of flavors, and novel compositions—knows no bounds.

Gin is based on natural ethanol; juniper berries are added and distilled with water to lower the proof and make the flavors stand out. While alcohol is necessary for capturing aromas, water is vital for bringing them to full expression. Now that juniper berries are no longer deemed sufficiently interesting, recipes for gin have become a veritable playground, with ingredients including apricot, peach, pear, red fruit (strawberry, raspberry), exotic fruit (passion fruit, mango, papaya, pineapple), citrus (lemon, grapefruit, orange), assorted flowers (jasmine, orange blossom, rose), aromatic herbs (rosemary, angelica, immortelle), bark (cinnamon), and spices (pepper, cardamom)—whatever serves to enchant the nose and palate.

Hardly a month goes by without a new brand appearing on the market. France is one of the world's leading consumers of gin, but the Philippines lead the way, followed by Spain, the United States, and Britain.

Gin gave its name to Gin-Fizz by Lubin (1955), a perfumed relaunched in 2009. It can also be found in Polo by Ralph Lauren (1978), Xeryus by Givenchy (1986), Kenzo Pour Homme by Kenzo (1991), and Voyage by Hermès (2010).

CAUTION

Juniperus communis should not be confused with cade juniper, or *Juniperus oxycedrus*, a fragrant, rot-resistant wood used for carpentry, door lintels, and moth-proofing. The difference between the two plants can be seen in their foliage. The needles of common juniper have a single white stripe on their upper surface, while cade juniper needles have two stripes. Cade oil is extracted by pyrolysis, or thermal decomposition of the wood, and is used as an antifungicide in shampoos. In France, Bébé Cadum soap owes its name to the substance, although it subsequently proved to irritate children's skin and was removed. Cade oil should not be confused with steam-extracted phenol-free essential oil, which has a scent similar to that of white birch.

Lower part of a plant; usually immersed in the soil, it always grows in the direction opposite to the stem and serves both to anchor the plant in the ground and to supply its food.
Littré Dictionary

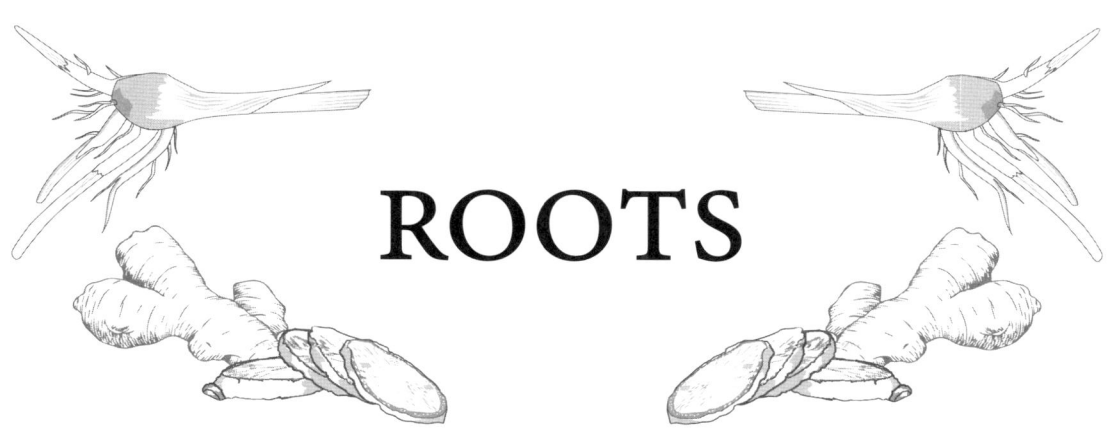

ROOTS

Whether employed for medicinal, magical, or erotic purposes, roots have uses that are confused and intertwined; most often, they are linked to popular beliefs handed down by word of mouth from one generation to the next. Thus, decoctions of vetiver root are reputed to stimulate the libido, while those of ginseng are supposed to enhance cognitive performance, attention, and concentration. We should recall the therapeutic and magical properties formerly attributed to mandrake. In the Middle Ages, mandrake collectors had to plug their ears to avoid hearing the plant's screams as it was pulled from the ground; otherwise, they faced a brutal end. In our own times, the adventures of Harry Potter have lent new life to the legend.

CALAMUS

Acorus calamus

Calamus is the Latin designation for a plant known as fragrant rush among botanists and perfumers. It's also the name for the pointed reed that gave rise to cuneiform writing when it was pressed into soft clay. Dipped in ink, calamus was subsequently applied to papyrus, parchment, and, finally, paper, paving the way for the quills that came later in history. This elegant aquatic plant can grow up to one meter tall. Its slender leaves resemble those of water irises, except they are thinner. Flowers are hermaphroditic, yellowish, lacking a corolla, and clustered in a lateral spike six to eight centimeters long. The rhizome is thick, branched, and highly fragrant, hence the plant's common name. Native to Asia, calamus grows in marshy areas and protects riverbanks from erosion. It is found in Korea, China, Japan, Russia, and North America, where it is also cultivated. The lower parts of young shoots are edible, but rhizomes are the most useful.

Before playing a role in perfumery, calamus was used in traditional Indian and Chinese medicine for its numerous virtues; it is one of those plants that count as a panacea. Additionally, its crushed and macerated leaves and rhizomes are effective against aphids, mosquitoes, and leafhoppers. It's the equivalent of nettle slurry, a natural insecticide that has been banned in France since 2006 but is still on sale.

Along with saffron, juniper, myrrh, and cardamom, ancient Egyptians used fragrant rush in the preparation of *kyphi*, their most common perfume. It is said that when Tutankhamen's tomb was opened, this is the scent that emerged. It's a fine story, but a myth: no scent, no odor molecule, can last for thousands of years. Calamus essential oil is distilled from dried rhizomes, with yields ranging from 2 to 5 percent, depending on root quality. Its complex scent is difficult to classify, somewhere between the smell of supple leather and the flavor of a pastry.

Calamus is used in trace amounts in perfume because of its unique character. European legislation limits its presence to a concentration of 0.02 percent. It is found in the composition of Acteur by Azzaro (1989), Calamus from the Olfactory Library collection by Comme des Garçons (2000), Un Jardin sur le Nil by Hermès (2006), and Eau Duelle by Dyptique (2010).

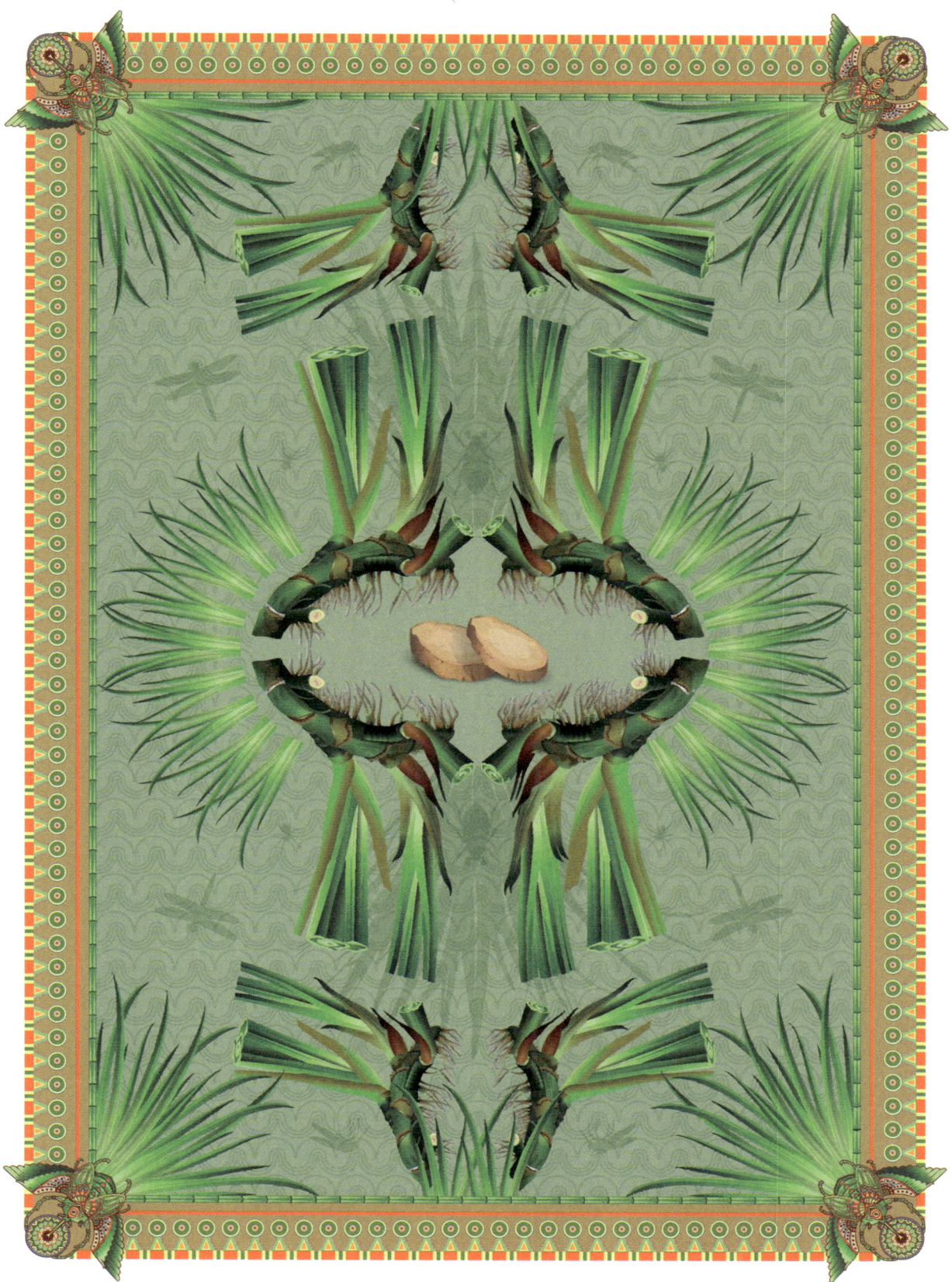

CYPRIOL OR NAGARMOTHA

Cyperus scariosus

"Mix nagarmotha oil with a lover's saliva, then menstrual blood, Indian spikenard, jatamansi rhizome, powdered human skull, and costus root. Apply a few drops to a man's brow and he will be sure to be happy in love."

This remarkable formula numbers among recipes belonging to the art of *Vashikaran*, an ancient Indian ritual practice based on magic charms that offered various ways to kindle love and sexual relations—long before the scientific discovery of pheromones. While they work well for invertebrates (insects cannot resist them), pheromones do not prove so effective among vertebrates. Our receptors for these molecules atrophied long ago; for ages now, humans have had a clear bias for visual communication.

I was shocked, stunned, and taken aback by what I had read. This recipe for black magic or witchcraft seemed to deserve a place in *Indiana Jones and the Temple of Doom*. Once my astonishment had passed, I gained insight into other ways of using smells: nagarmotha oil represents a means of subjugating one's partner by force. It makes sense that such a spell would come from India. This is the land of the *Kamasutra*, an ancient Hindu text dealing with private life, including erotic practices. Perfumes have existed for ages in India in the form of attars: co-distillations of plants, wood, roots, spices, flowers, and leaves with a fatty substance (usually sandalwood oil) that serve to perfume the body without the presence of alcohol.

Cypriol is an herbaceous plant with a dense root system. In fact, it's so dense that the plant has been called the worst weed in the world and, like papyrus on the banks of the Nile, overly invasive. Roots are harvested from February to June by villagers and transported to a factory where they are washed and dried before distillation; vetiver roots are processed in the same way. It takes 250 kilograms of dry roots to produce one kilogram of essential oil, which has an earthy, animal smell reminiscent of castoreum. Its reasonable cost helps make it a favorite for crafting oud perfumes. Owing to its complex composition, cypriol is difficult to identify formally in a perfume; one has to rely on the description of fragrances used—or the nose of a professional.

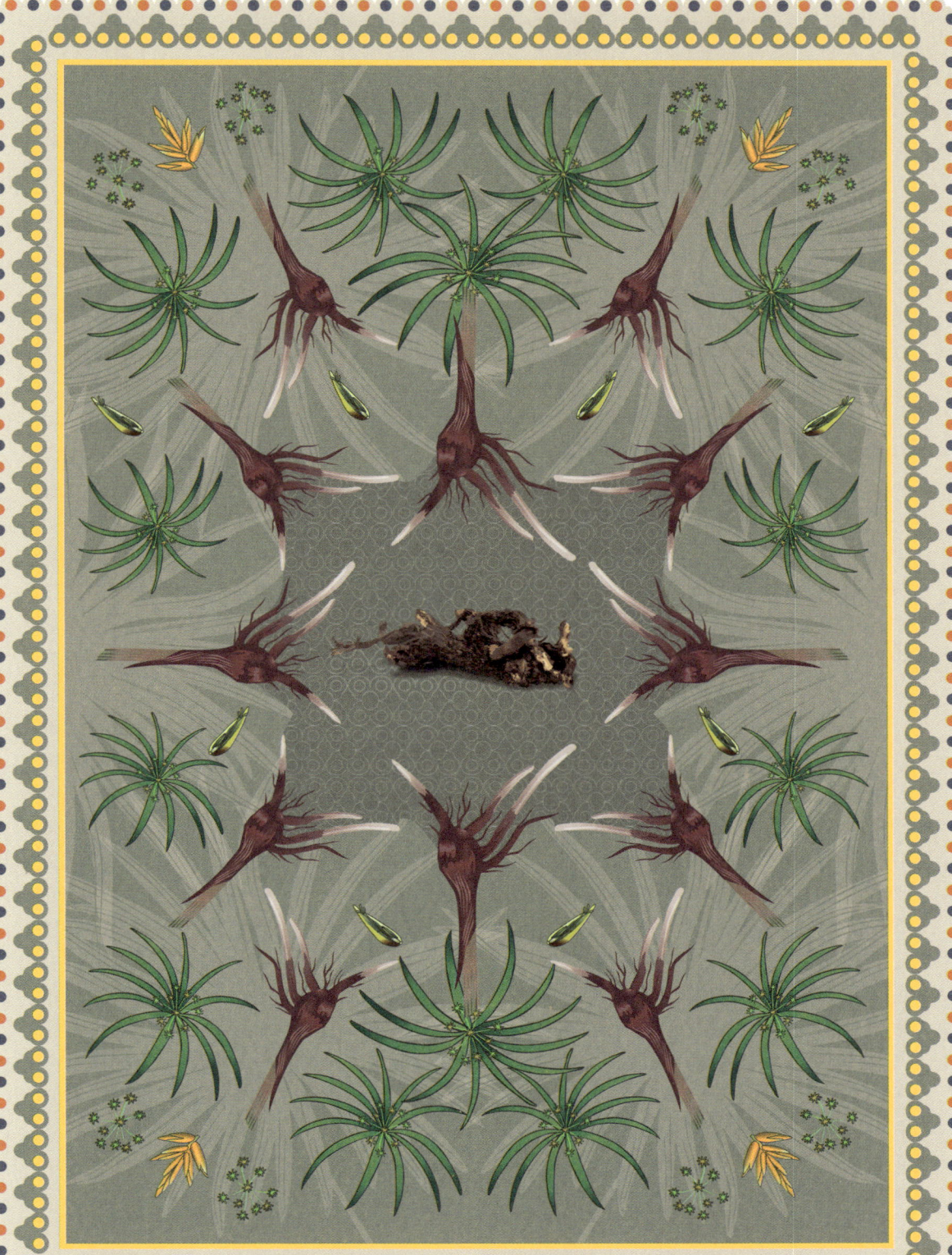

GINGER

Zingiber officinale

Native to India, ginger is a tropical and ornamental perennial that can reach up to 1.5 meters. It grows in India, China, Jamaica, Africa (Ivory Coast), along the Mediterranean rim, and even in pots at home; it prefers indoors, as long as conditions are warm (25°C) and its soil slightly moist. The leaves are evergreen, lanceolate, and long. The white flowers have a scent that evokes *Hedychiums*, which are plants with very similar rhizomes, and in particular *Hedychium flavum*, which smells of spicy jasmine.

Ginger first came to the West in the days of ancient Greece and Rome. From the first century on, it spread throughout Europe thanks to Arab merchants. Commercial centers on the Malabar coast, especially the port of Cochin, exported spices to the Mediterranean world. In the account of his journey to India, Marco Polo mentions how shippers took advantage of the cycle of monsoon winds to trade their wares. Spices from many other parts of Asia also came this way. Today, ginger is the most widely used spice in the world; three million tons are consumed annually. The second most popular spice is black pepper (*Piper nigrum*), at 350,000 tons.

The rhizome is the vital part of the plant for both food and perfumery. Ginger root contributes to curry in India, serves as an accompaniment to sushi in Japan, and features throughout traditional cuisine in China. Closer to home, it is used to make ginger ale, a beverage the Irish are said to have invented in the eighteenth century. Its essential oil has long been extracted from dried rhizomes. It was not until the 1990s that perfumers were offered a fresh ginger essential oil, from the Ivory Coast. The yield is around 2 percent. The main odorous constituents are citral, neral, and geranial, which also occur in lemon and verbena. Ginger has one of the brightest and most shimmering fragrances in the palette of scents; it is often combined with cardamom, whose essential oil is obtained from a seed belonging to the same botanical family, *Zingiberaceae*. Ginger is featured in perfumes including Cartier's Declaration (1998), Hermès's Un Jardin Après la Mousson (2008), Yves Saint-Laurent's L'Homme Cologne Ginger (2011), and, more recently, Hermès's Twilly (2017) and Twilly, Eau de Ginger (2021).

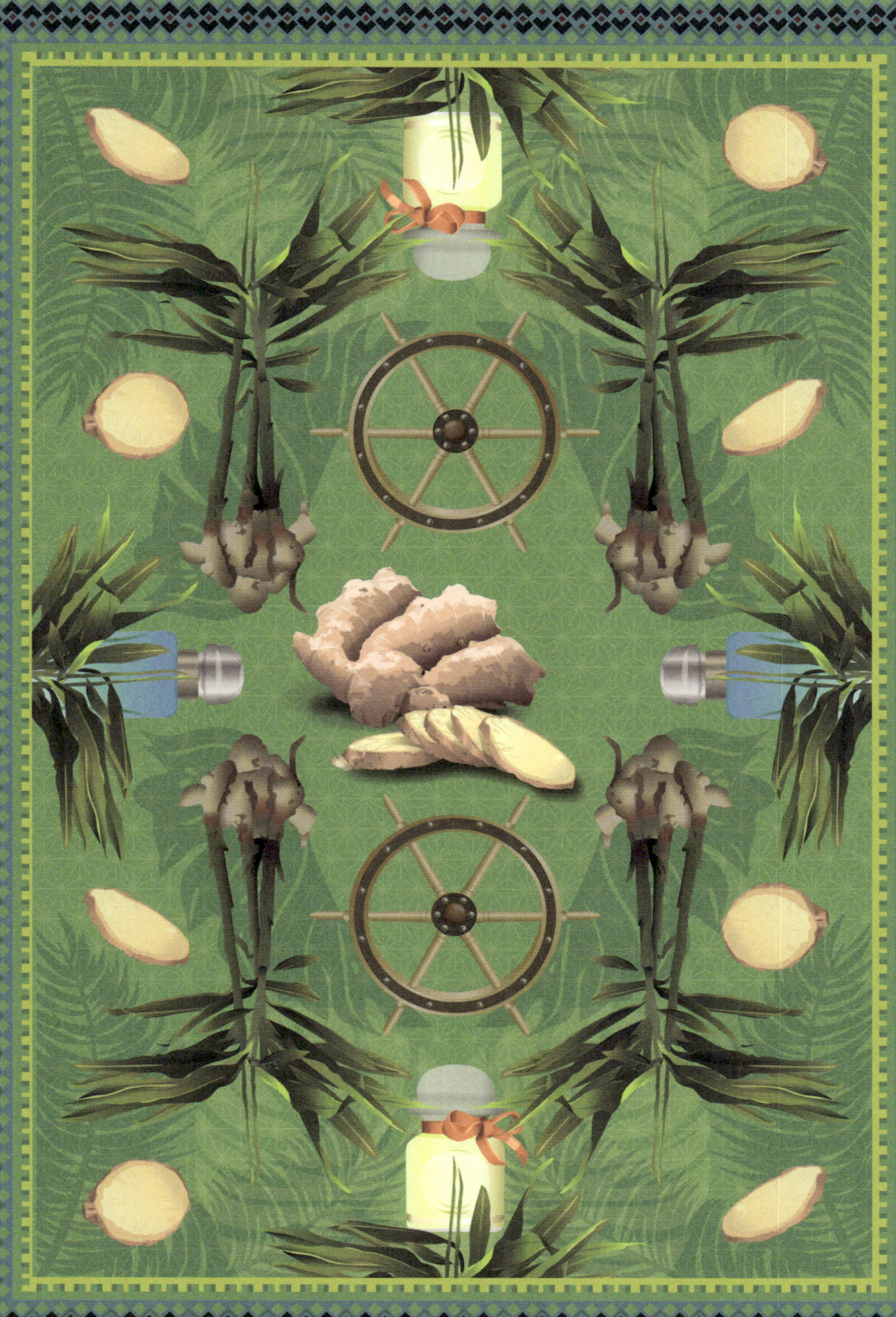

JATAMANSI OR SPIKENARD

Nardostachys grandiflora

To speak of spikenard is to speak of one of the oldest known perfumes. In the Gospel of John, it is written: "Six days before the Passover, Jesus came to Bethany, the home of Lazarus, whom he had raised from the dead. A meal was given in honor of Jesus there. . . . Mary took a pound of very pure and valuable spikenard; she anointed Jesus' feet and wiped them with her hair. The house was filled with the fragrance of the perfume. Judas . . . one of his disciples, then said: 'Why was this ointment not sold for three hundred pieces of silver, and the money given to the poor?' . . .

Its fragrance wavers between the earthiness of vetiver and the humus of patchouli

Jesus said to him, 'She has kept this perfume in preparation for the day of my burial. You will always have the poor with you, but you will not always have me.'"

Jatamansi, which is native to China, Bhutan, India, and Nepal, goes by many names: Indian spikenard, Himalayan spikenard, Indian valerian, and Bengal valerian. The plurality of designations shows the importance given to this herbaceous perennial of the *Valerianaceae* family; in his *Natural History*, Pliny the Elder already counted a dozen species of spikenard with similar properties and characteristics. The plant has lanceolate leaves from which a stalk emerges bearing small flowers of a pretty mauve in one or more clusters. Steam distillation of rhizomes produces the extract. Although jatamansi belongs to the same family as European valerian, it does not have the latter's brutality. Its fragrance wavers between the earthiness of vetiver and the humus of patchouli. This highly coveted root has been added to the CITES list. As a result, imports to the European Union from Nepal, an important supplier, have been halted for a few years now.

The scent expresses itself in Jatamansi by L'Artisan parfumeur (2007). It joins cypriol to heighten the interest of Oud Élixir Précieuse by Dior (2014) and Oud Absolu by Cartier (2016).

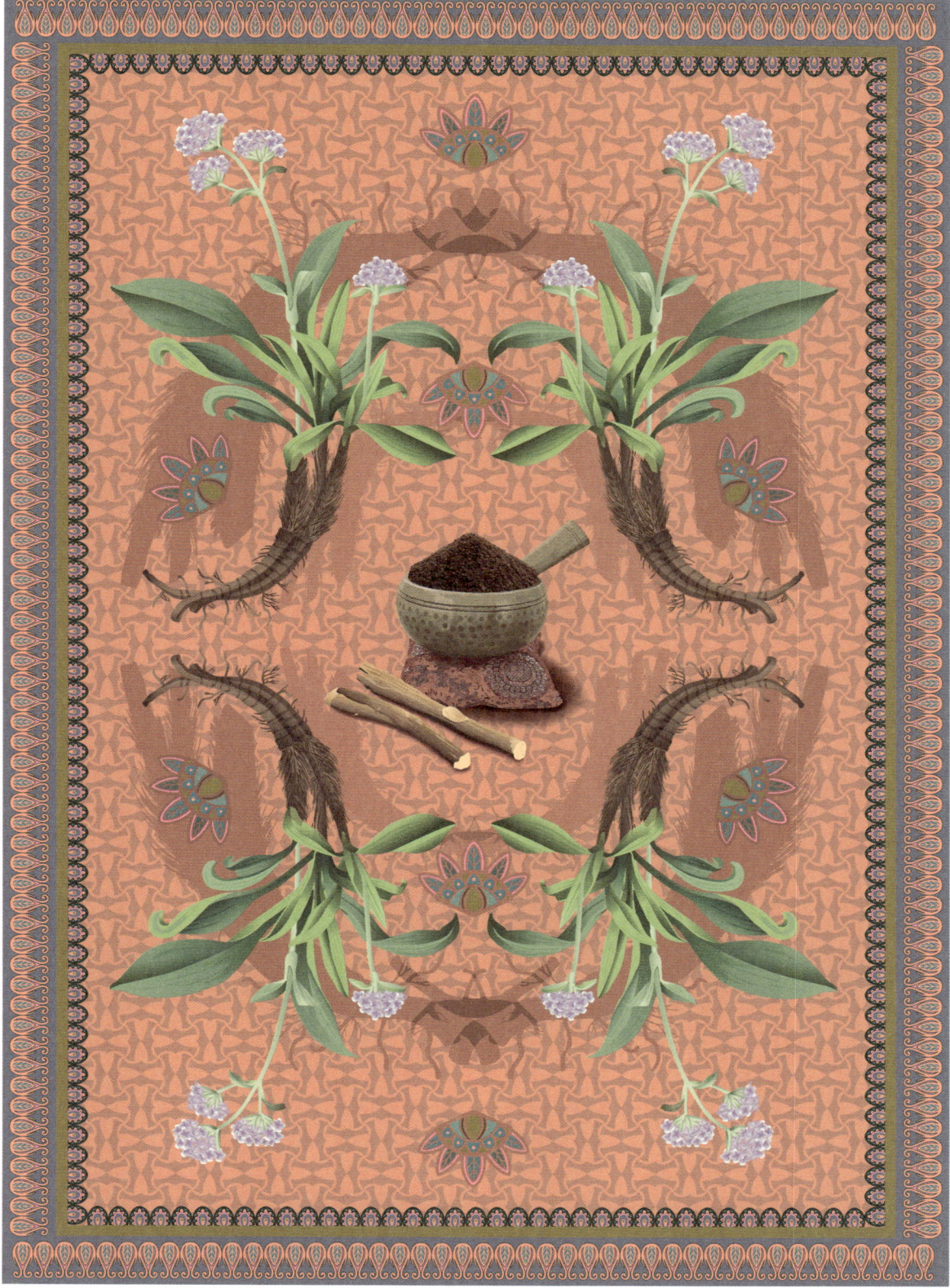

ANIMAL PRODUCTS

The subject of this book is botany and its relationship to the world of scents and perfumes. Our discussion would not be complete if it did not include animal products, whose smells are also found in the plant world. In springtime, my garden boasts a charming orchid called *Himantoglossum hirsinum*, which has the odor of goats, and poet's daffodil (*Narcissus poeticus*), whose smell recalls horse dung. In order to be fertilized, flowers emit scents of animals, feces, and even sperm. A letter penned by the Marquis de Sade mentions the latter: "The chestnut flower positively has the same odor as that prolific seed which it has pleased nature to place in the loins of man to reproduce his kind." The animal-derived ingredients in perfumery are beeswax, castoreum and civet absolutes, musk and ambergris (which are employed in the form of an infusion), and finally, ambrettolide, a molecule related to a synthetic musk of animal origin.

CAUTION

No current legislation prohibits the use of these materials. Musk deer are protected by CITES, but musk is still permitted. If perfumers are now abandoning these substances, it is because of their high cost and rarity; even more, a tacit code of ethics has prompted members of the profession to reconstruct them by artificial means.

BEESWAX

"What color stripe do we start the back of the bee with?" asked one of my grandsons, who was coloring a drawing he had made. I replied: "I don't know, you can use whatever color you want." I promptly regretted my hasty response: it closed down the conversation. My grandson's question testified to his curiosity and concern for the truth. Together, we could have found an answer.

We love bees. When they get too close and buzz around us, we get nervous, but all the same we are ready to take to the streets to demonstrate against pest-control products that disorient and kill them. If the pollination of flowers, fruits, and vegetables is at stake, then our own sustenance is also on the line.

The beeswax used in perfumery comes from the cracks of old frames in the upper part of hives, which honey producers consider waste. Black with wear and full of scents, they represent a wonderful by-product for procurers of raw materials: they're easy to find and can be used in a circular economy.

Beeswax absolute, with its characteristic smell of honey, is obtained through extraction first with volatile solvents and then with ethanol.

CASTOREUM

In *Don Quixote* (1605), Miguel de Cervantes has his hero observe that "the pagan acted sensibly and imitated the beaver, which, finding himself hemmed in by hunters, bites himself and rips away with his teeth the thing that he knows by instinct he is being hunted for." As these lines reveal, it's not just the animal's fur that is prized. In Canada, the large rodent is considered a forest engineer because it maintains wetlands and waterways; a few naysayers deem it a destroyer of these same forests. Both females and males have a large pair of glands under the tail that produce an oily secretion called castoreum; they rub themselves with this substance and use it to mark territory for the benefit of members of their clan. Recovered from animals that have been raised or hunted for their fur, the glands are dried and sold to perfumers to extract a resinoid or absolute that has a powerful, phenolic smell of leather, similar to white birch essence.

CIVET

When I was young, the company I worked for gave the order to no longer use secretions from the anal gland of civets, as the substance was obtained by means of curettage. I protested vigorously. I knew that farm-raised animals represented an important source of income for the impoverished people of the Horn of Africa, and that families exercised the greatest care in tending this valuable resource. In the 1990s, the perfume industry, which is always attuned to the beating of bleeding hearts, called for an end to the use of civet. A few years later, a representative of a global organization called to tell me that he had heard about my objections, and that families had struggled when their income abruptly stopped. He asked me how his organization could compensate them. I lost my temper again and told him that we shouldn't make decisions based on our own, Western navel-gazing. The damage was done. Since then, we have been using a synthetic civet. The lure of profit had won. Civet is an expensive substance that has been used for millennia to enhance floral notes in perfumes. In the Middle Ages, it served to enhance the gaminess of dishes, which shouldn't be surprising: people loved meat diets then.

NATURAL MUSK

Moschus moschiferus is the Latin name for the Siberian musk deer. During the rutting season, the animal's abdominal gland, which is the size of a chestnut, secretes a powerful odorous substance with a smell similar to human blood. This is a source of both appeal and repulsion.

Deemed a panacea, elixir, philter, and aphrodisiac that can heal, conjure, bewitch, and charm, musk has been prized for more than 2,000 years in Asia. In China, ink was perfumed with musk grains in order to under-score the value of texts and documents. In Japan, samurai headed into battle carried small leather pouches holding small quantities that they chewed to give themselves courage. In the West, musk arrived on the scene during the French Revolution. Young royalists, who were very elegant, applied it so copiously that they came to be known as *muscadins*. At the beginning of the twentieth century, musk found use only in the form of an infusion. In our own time, to protect the species, hunting musk deer is prohibited. Farms have been established, and exports are subject to very specific quotas supervised by CITES. The use of deer musk (or Tonkin musk) is authorized, then. Synthetic musks, which are musks in name only, cannot replace them; they are different products with different smells, and they do not bind scents. They last for a long, long time.

AMBERGRIS

"Though the word ambergris is but the French compound for grey amber, yet the two substances are quite distinct. . . . Ambergris is never found except upon the sea. . . . Amber is a hard, transparent, brittle, odorless substance, used for mouth-pieces to pipes, for beads and ornaments; but ambergris is soft, waxy, and so highly fragrant and spicy, that it is largely used in perfumery, in pastiles, precious candles, hair-powders, and pomatum. The Turks use it in cooking, and also carry it to Mecca, for the same purpose that frankincense is carried to St. Peter's in Rome. Some wine merchants drop a few grains into claret, to flavor it."

When he devoted a chapter of *Moby-Dick* to this substance, Herman Melville sought to bring some order to the vast and confusing world of whalers. He tells how some men were tempted to hunt sperm whales in order to collect ambergris, which was most often found floating at sea or deposited by currents on coasts as far away as Ireland. This intestinal lithiasis, which evidently results from the interaction of bile secretions and food the whale has ingested, has assumed mythical stature. It gives perfumes an untold, mystical dimension, capable of rousing infinite fantasy. Its smell of cold cigarette ash has always seduced me—so much so that I incorporated an infusion of it, at 10 percent, in L'Or Black by Pascal Morabito (1981), a scent for men that has now disappeared.

AMBRETTOLIDE

Ambrettolide is a musk obtained from aleuritic acid, which comes from the substance secreted on the bark of trees by *Kerria Lacca*, an Asian insect in the *Kerriidae* family. This resin plays a role in making cosmetics and varnishes. In the early days of the recording industry, it was used to manufacture phonograph and 78 rpm gramophone records. In 1948, vinyl, a plastic material derived from ethylene, took its place. Ethylene, for its part, has been used to produce synthetic musks. A vegan version of ambrettolide exists, which is made from cane sugar.

GLOSSARY

Absolute : Product obtained when parts of a concrete soluble in ethanol have been cold-extracted.

Concrete : Product obtained from fresh plants (leaves, flowers, lichens, seeds, wood, etc.) by extraction with a volatile solvent (hexane, carbon dioxide).

Cultivar: Variety of a plant species selected for cultivation.

Essence : Product obtained only by cold expression of citrus peels.

Essential oil : Product obtained by steam distillation of fresh or dried plants.

Resinoid : Product obtained by ethanol extraction of balms and resins, followed by evaporation of the alcohol.

Traces : Fragrant product used in perfumes in minute dosages.

CLASSIFICATION

CLASSIFICATION OF THE NATURAL MATERIALS LISTED IN THE TWO ATLASES AND USED IN PERFUMES

* Unless otherwise indicated, the natural products mentioned are mainly essential oils.

** abs. = absolute

Woods and Barks

Cedarwood: Virginia, Alaska, Texas; Sandalwood: India and Australia; Rosewood; White birch; Guaiac wood; Ceylon and Chinese cinnamon; Cypress; Fir balsam (abs.); Oak moss and Tree moss (abs.); Oud and Boya

Leaves

Absinthe and Wormwood; Basil and Tarragon; Cistus and Labdanum (abs.); Davana; Hay (abs.); Bourbon Geranium; Mastic (abs.); Lovage: leaves and roots; Mints: Peppermint, Crepemint, American mint, Eau de Cologne mint; Patchouli; Petitgrain; Rosemary; Tea and Mate (abs.); Thyme; Violet (abs.)

Flowers

Indian and Sambac Jasmine (abs.); Lavender; Lavandin; Magnolia flowers and leaves; Narcissus (abs.), Orange (abs.), and Neroli; Osmanthus (abs.); Turkish Rose and *Rose de Mai* (abs.); Tuberose (abs.); Ylang-Ylang; Acacia (abs.); Broom (abs.); Tagetes; Immortelle (abs.); Roman and Blue Chamomile

Fruits

Bergamot; Blackcurrant; Lemon; Mandarin; Sweet and bitter orange (Bigarade); Citron; Lime; Tangor; Yuzu

Gums and Resins

Benzoin; Frankincense and Myrrh; Opopanax; Galbanum; Elemi; Balm of Peru and Tolu

Seeds

Amber; Cardamom; Carrot; Celery; Cumin; Clove; Mace and Nutmeg; Black pepper; Sichuan pepper; Timut pepper; Pink berries; Tonka (abs.); Vanilla; Star anise; Coriander; Juniper

Roots

Angelica; Iris (concrete); Vetiver: Haiti and India; Calamus; Cypriol; Ginger; Jatamansi

BIBLIOGRAPHY

Actes des colloques du M.I.P. *Un jour une Plante.* Studio VéAche, 2008.

Arctander, Steffen. *Perfume and Flavor Materials of Natural Origin.* Elisabeth, NJ, 1960.

La Fabuleuse histoire de l'eau de Cologne. Nez Éditions, 2019.

Maret, Léon. *Les Fleurs de la Côte d'Azur.* Paul Lechevalier Éditeur, 1926.

ACKNOWLEDGMENTS

Société Raices Verdes: for the rosewood

Société Biolande: M. Renaud Beguin de Billecoq

Société Floral Concept: M. Frédéric Remy

Société Simone et Gatto: M. Villefedro Raimo

M. Stéphane Piquart: Sourceurs de Matières premières